Broadcast Television
Effects in a
Remote Community

LEA's Communication Series

Jennings Bryant and Dolf Zillmann, General Editors

Selected titles in Mass Communication (Alan Rubin, Advisory Editor) include:

Bryant/Bryant • Television and the American Family, Second Edition

Harris • A Cognitive Psychology of Mass Communication, Third Edition

Perse • Media Effects and Society

Van Evra • Television and Child Development, Second Edition

Wicks • Understanding Audiences: Learning to Use the Media Constructively

For a complete list of titles in LEA's Communication Series please contact Lawrence Erlbaum Associates, Publishers.

Broadcast Television Effects in a Remote Community

Edited by

Tony Charlton
*Cheltenham and Gloucester College
of Higher Education*

Barrie Gunter
University of Sheffield

Andrew Hannan
University of Plymouth

LAWRENCE ERLBAUM ASSOCIATES, PUBLISHERS
2002   Mahwah, New Jersey   London

Lawrence Erlbaum Associates, Inc., Publishers
10 Industrial Avenue
Mahwah, NJ 07430

Cover design by Kathryn Houghtaling Lacey

Library of Congress Cataloging-in-Publication Data

Broadcast television effects in a remote community / edited by Tony
   Charlton, Barrie Gunter, Andrew Hannan.
       p. cm. — (LEA's communication series)
     Includes bibliographical references and index.
   ISBN 0-8058-3735-3 (cloth : alk. paper) ✓
   1. Television and children—Satin Helena. 2. Television broadcasting—
     Social aspects—Saint Helena. I. Charlton, Tony. II. Gunter, Barrie.
     III. Hannan, Andrew, 1951– IV. Series.
   HQ784.T4 B76   2002
   302.23'45'09973 —dc21   2001051282
                                                                    CIP

Books published by Lawrence Erlbaum Associates are printed on acid-
free paper, and their bindings are chosen for strength and durability.

Printed in the United States of America
10 9 8 7 6 5 4 3 2 1

Contents

Foreword

Pamela Lawrence

In many ways, St. Helena is unique. People still make time to greet each other as they pass on the street and meet in shops. Moreover, most young children grow up wrapped in senses of belonging, security, and familiarity that—if not provided for by the immediate family—will come from the extended family (e.g., siblings, grandparents, aunts, and uncles). Other adults, particularly school teachers, also offer stability and security to youngsters whether they are in school, at play, or in the wider environment.

Within this small population everyone is connected with someone else, so most people are known to most others. Links of this kind foster a sense of camaraderie and security reinforced by a watchfulness in the community. This watchfulness, and how it influences people, can be illustrated in the school setting. It is not unusual for a pupil to have a relative or family friend as their teacher. On their way home from school, the teacher is likely to meet the parents or someone closely connected to the pupil's family (e.g., a grandparent or neighbor). School matters, including concerns, can comfortably be raised in their conversation so that the pupil's family quickly becomes aware if anything untoward has happened. Such links often act as a deterrent to unacceptable behavior in school.

On St. Helena "even the rocks have eyes and the winds capture your words and carry them to where you least want them to go." It is so easy for the local grapevine to spread news about people and events. There are positive and negative aspects to this awareness. When a tragedy occurs or an individual calls for help, news spreads quickly and people readily gather to provide support. This lack of anonymity and privacy works in other ways, too. People cannot hide from their responsibilities, so it is necessary to "survive" both the good and bad times. It means, also, facing responsibilities "head on." Family and community function as a mirror, providing a continuous reflection to people on their behavior. To avoid an unfavorable reflection, individuals have to "behave as expected," or face the consequences of "stepping out of line." Not everyone can cope with these "demands" and expectations. Some may not have what it takes to overcome the threat of self-fulfilling prophecies for negative expectations such as "It's not surprising he's gotten into trouble; his daddy is a trouble-maker, as was his granddad before him. It's in the family." There are others, too, who are overly conscious of having to "follow in the footsteps" of parents or other family members who have been not only "successful" in some way but also much respected because of that success. Islanders are often remembered not just in terms of their own behavior but also in terms of the deeds of their grandparents or even great-grandparents.

In a small community, sorrows and joys are shared. Empathy is easier when you know those who experience tragedy or happiness. Similarly faults and failings are "shared," as are successes. Students embarking on overseas study know they are expected to do well. They have to live up to family (as well as their own) expectations in addition to the community expectation. A failure to live up to these goals means having to "face" the whole community, whereas living up to positive expectations gains respect in the community. Most people "look up to" achievers and perceive them as positive role models.

Pressures of family and community expectations cannot be underestimated. There are the unwritten rules that, if broken, leave a price to be paid. These include the customs and traditions that are an integral part of the island's character and way of life. These range, for example, from the etiquette for dealing with births, deaths, and marriages to reacting to major disasters, the joys of homecomings, and raising a family, as well as eating and entertainment. The degree of tolerance afforded to those who break the "unwritten rules" is perhaps less than for those who break the "written" ones. A person can serve a jail sentence and then, depending on the circumstances, be given support and become rehabilitated within the community. In contrast, a quarrel over the preparations for a marriage reception (e.g., forgetting to include a relative on the guest list) could cause a lifelong rift between families or at least a simmering resentment.

St. Helena has recently experienced testing times, politically, economically, and socially. These experiences are already affecting island life. Traditional customs and routines are either disappearing or under threat. For example, cohabitation (as opposed to marriage) has become more acceptable in recent times. In turn, this practice may impact on the degree of commitment to the extended family. Although there is still a housing problem on St. Helena, young families are increasingly owning their homes. Consequently, although the extended family continues to live "just down the road," they are neither as accessible or influential in imparting traditional values to young families as they once were. Similarly, the role of grandparents is changing in line with the changing status of women. Many more women are working today. In past times, grandmothers and other relatives (especially if they shared the same dwelling) often cared for children of working mothers. Nowadays, this practice is diminishing; even grandmothers tend to be busy pursuing other work. This change is regrettable, for many families have already become fragmented because fathers (and now mothers) work offshore on Ascension Island, the Falkland Islands, the RMS *St. Helena* or in the UK.

St. Helena has a history of offshore fathers. Limited employment prospects compelled many fathers to work overseas. Today there are many instances where both parents are working offshore. Sometimes the absence of the parents appears to adversely affect children's behavior. This may arise from a lack of security and stability within the surrogate home. When grandparents or close relatives cared for children, fewer problems were apparent. This unsettling situation may help explain the apparent increased "restlessness" within some of the children at school.

Changing work practices affect children in other ways. Rather than returning home for lunch, most children now remain at school. So the main meal of the day is in the evening. This mealtime gathering allowed families to come together to talk. Anecdotal evidence suggests that on St. Helena the family gathering at mealtime, like in the UK, is on the decline. (Statistics show that only 9% of families in the UK have a family meal and 24% do not have a table around which the family can sit.)

Church attendance and religion are in decline as well, although there is still a quiet religious "spirit" around the island. This is clearly visible when death threatens or disaster strikes. When someone is seriously ill, families call for religious support; when someone dies, a religious burial is arranged. More generally when disaster falls, it is shared by the entire island community.

Leisure time is another changing aspect of life. The warm climate draws people outdoors. Even so, Jamestown people seldom sit on their doorsteps as much as they used to. I recall childhood memories of sitting with friends by a neighbor's door in the early evening or on a Sunday afternoon. We talked with each other, held sing-songs while others played guitars or played games. There are still some times when people can be seen sitting outside, calling out to passers by, and talking to friends who stop. This "open interaction" is not as prominent in today's St. Helena for many reasons. Gone are the days when families shared domestic facilities like a yard or a communal water tap. Shared living and working spaces helped create bonds between neighbors. Housing accommodation is now more sophisticated and many families live in independent units behind "closed doors." The new homes built outside of and above Jamestown are detached, consequently it is often only by chance encounters that people become involved in each others' routines. Furthermore, the telephone has diminished the need for stepping out to call on family or friends.

Other events and institutions have outlived their purpose or value. Friendly benefit societies have declined in membership and, consequently, the annual anniversary "March Days" have been reduced from 5 to just 2 (Children's Society and the Church Providence Society for Women). Not long ago these March Days were among the highlights of the year, and for many they were among the rare occasions when people visited Jamestown from surrounding districts to join the festivities.

The traditional New Years Day School Sports and Treat Day no longer occurs. Nowadays, children do not rely on special occasions for treats. The enthusiasm for the traditional school sports has declined. People prefer having time to celebrate on New Year's Eve and to socialize at leisure on New Year's Day. In recent years, roadside and playground games have also lost their seasonal appeal. Once there were particular times of the year when hopscotch, five-stones, skipping, tic-tac-toe, and other games would be in season. Various explanations help account for neglecting to include play sites in new housing areas—the popularity of videoviewing and broadcast television viewing are among these explanations.

Respect is changing too. There is still a strong desire, particularly among the older generations, to please those in "positions of power." In a community that has been raised on subservience, and giving way to authority and others with elevated status, this is hardly surprising. There is a difference, however, between enforced subservience and genuine respect for authority, for family elders, older people, and people in general. There are still many people who automatically address their elders or those in authority as "Ma'am" or "Sir." It is still difficult for some senior colleagues (e.g., in schools) to address an older counterpart by their Christian name, even when best of friends.

There are many explanations for this apparent change. Adults now have to "earn" the respect that they once received automatically (because of respect for age). From a school perspective, a sense of assertiveness is developing among the younger generation. This can be positive for St. Helena, provided it does not lead to a divestment of genuine respect, good manners, and family cohesion. A greater awareness of the outside world mainly through exposure to television and improved education as well as through access to a range of new electronic technologies (e.g., the Internet), are all factors that influence the aspirations and opinions of island youth. There is a paradox at work here. Assertiveness, when positively channeled, can help to fire the aspirations of the younger generation and help raise the profile of St. Helenians. But is family cohesion, social stability, and economic promise strong enough to meet these aspirations, or will frustration lead to negative consequences (e.g., through unfulfilled potential)? Will future family and community values be strong enough to counteract the negative consequences of exposure to the "outside"? Only time will tell.

On the island the feeling of "belonging" still exists in its broadest sense. This is evident on occasions such as Christmas, which occurs in the southern hemisphere's summer. The festivities have been described by outsiders as "one long party." Families visit friends and neighbors. Each Christmas morning in Jamestown, for instance, people group together and visit homes to extend greetings; they have a drink and a sing-along before moving on to the next house. The groups grow in numbers as they move along. The atmosphere is one to be savored.

All customs are not lost; there is still much to look forward to in events such as the Maundy Thursday fishing boat parties. Jamestown Wharf is also filled with spectators and people fishing. On Good Friday, coffee and buns form the traditional breakfast and fresh fish is eaten instead of meat. During Easter, large groups of families and friends camp together in the country at Bluehill, Thompson Hill, and Horse Pasture Plain or around the rocky coastline at Lemon Valley or Rupert's Bay.

Ship Day is another occasion when crowds gather at the wharf either to greet family and friends returning from overseas or just to see who is arriving. When the ship is scheduled for departure, the crowds gather once again to bid farewell to family, friends, or departing visitors. These are the occasions in life when islanders get caught up in the atmosphere—the joy of homecomings and the sadness of farewells. Such matters are part of a St. Helenian's heritage. The belief is as strong as ever that "to better oneself, one has to leave the island." Although the island has so much to give in the way of quality of life, many still feel the need to leave and appreciate this quality from afar. Absence makes the heart grow stronger.

How a person performs or behaves as a St. Helenian living in St. Helena is very much attributed to the uniqueness of life in the community as described. What makes a Saint who she or he is, what motivates or demotivates her or him, whether she or he excels in meeting life's challenges is largely a product of all that happens on the island.

The following chapters are about islanders; about the lives of the people as much as they are about the ways in which the people respond to TV. If findings to date are correct, they tell a great deal about the quality of family and community life on this remote island in the South Atlantic Ocean.

Preface

This volume reports findings from a major, multidisciplinary study of the impact of broadcast television on the remote island community of St. Helena in the South Atlantic Ocean. Broadcast television was introduced to the island for the first time on March 31, 1995 when a single television service, Cable News Network (CNN), was beamed in by satellite. In the following months, further services were added to a two-channel system including movie, documentary, sports, and children's services. Viewing quickly became available around the clock. More recent changes to the local television network introduced a mix of KTV, Movie Magic, BBC, Discovery, and Supersport.

The introduction of broadcast television in this way represented a major event on the island, whose inhabitants' only televisual experience had been through video. The population enjoyed a limited experience of cinema viewing and earlier encounters with magic lantern shows. The additional televisual experience introduced an intervention in the local environment that set in place an opportunity for a quasi-experimental investigation to be undertaken into the impact of broadcast television on social behavior in terms of content and displacement effects. This opportunity was seized by Tony Charlton, who obtained permission to instigate a research project on the island to investigate the effects of television, especially on the island's children. Funding from *The Sunday Times*, and later the UK Economic and Social Research Council in 1997, allowed the continuation and then the expansion of the research project.

The St. Helena project team began collecting data in November 1992, more than 2 years before the inception of broadcast television, to establish baseline measures on key variables that would feature as central aspects of the study. These included measures of how children on the island spent their leisure time, measures of their classroom (on-task levels) and around school social behavior, as well as their interactions in free-play settings in playgrounds. This meant that significant evidence was amassed on the status of young children's behavior patterns in an environment devoid of broadcast television and, indeed, with only limited other audiovisual experiences.

The project has continued since that time, and data have been collected annually from the island on the key dependent measures just mentioned. In addition, since 1995 data have also been collected on children's television viewing behavior, parental mediation of their viewing, older students' perceptions of television's influence, and the nature of the television output. The St. Helena project has, therefore, provided a rare opportunity to study in depth a series of relationships between children's experience of television and different aspects of their behavior.

Although this is not the first project of its kind, it has attempted to learn from previous studies that have been conducted in the United Kingdom, United States, Japan, Canada, and South Africa. In so doing, the St. Helena project has adopted a multimethodological and multidisciplinary approach to data collection. Data have been collected through quantitative and qualitative research techniques. The research team comprises specialists from the fields of communications, sociology, education, anthropology, and psychology. Data have been obtained at aggregated group levels and from individuals. Self-report data from the children have been supplemented by data provided by older students and teachers. Thus, a number of distinct perspectives have been integrated to yield a rounded understanding of the impact that broadcast television has had.

Pamela Lawrence's Foreword sets the scene for the chapters that follow. She provides a cursory, yet important, picture of life on the island. In chapter 1, the St. Helena Research Project is introduced. It reflects on the genesis of the project and outlines the various substudies that are part of the wider research study. Chapter 2 focuses on young students' written responses to an essay entitled "People on St. Helena." Students' remarks are discussed in the context of particular themes that are linked broadly to "social and community relations on St. Helena." In chapters 3, 4, 5, and 6 findings from the major substudies are reported. The research completed thus far has examined the nature of television output on St. Helena and sought evidence for possible effects of broadcast television on youngsters' leisure-time activities and the social conduct of younger children. Chapter

7 considers directions in which future research on the project can usefully proceed.

From an academic research standpoint, this volume presents a detailed example of the importance of naturalistic fieldwork in gathering knowledge about, and increasing our understanding of, the role and effects of a mass medium such as television. Much research into television's purported effects has been conducted either in artificial environments or in everyday environments where many uncontrollable factors exist (in addition to television) that can affect audience behavior; the current study represents an investigation that takes the form of an "experiment" in many ways but one that does not occur within the confines of a laboratory. The "laboratory" in this case is the natural (yet remote) living environment of St. Helena's population. This is not to say that this type of research is perfect in every way; it is not. There are still problems concerned with the control of extraneous variables. There are also risks associated with requesting too much of respondents in terms of providing data and even with observing their behavior. These issues are discussed here.

Finally, the research study is an ongoing project. Consequently, new data arrive throughout each year. The studies presented here do not always report on the most recent data collections. For example, in chapter 6 Tony Charlton and colleagues consider post-TV data collections from 1998. Other data for 1999 had arrived yet—although coded and analyzed—there was insufficient time to include it in the chapter. So the content of this volume reports only interim findings from what is to become a much lengthier longitudinal panel study.

ACKNOWLEDGMENTS

We owe thanks to everyone who made this volume possible. To Tessa Lovemore, Angela Hodgkinson, David Lovemore, Carol Lambden, Charlie Panting, Saskia Mansfield, Cilla Thomas (when studying in the United Kingdom), Kate Hodgkinson, Steve Treble, Ruth Bullivant, Carol Jones, and Shirley Mawer for the several hundred hours they spent coding television programs' content and children's actions in video recordings of playground behavior, and to Karen Somerfield who undertook the mammoth task of coding and data entry for the displacement study. Their dedication to this work was remarkable. We were fortunate to be able to call on the help of such colleagues.

To Charlie Panting for her Herculean administrative efforts in preparing the manuscript to send to Lawrence Erlbaum Associates. To Kenneth David, Dorothy Evans, Trevor Hearl, and Alexander Schulenburg, who read through the chapters to verify historical and other accuracies. The book became more scholarly as a consequence of their comments. We are indebted too to Alexander Schulenburg for the work he undertook on our behalf in the Castle archives on St. Helena.

To *The Sunday Times* for some much needed financial support in the early days of the project, to the Economic and Social Research Council (R000222070) for providing the major source of funding for the project during the last 3 years, and to Mick Abrahams, who somehow always managed to locate extra funds in our times of need.

To Taylor & Francis Limited (P.O. Box 25, Abingdon, Oxfordshire, England OX14 3UE) for permission to include in chapter 6 some material first published in *Research Papers in Education, 14*(3).

To St. Helena Island. It would not have been practicable to undertake the research project without the help of so many of our colleagues on St. Helena. In this respect we are grateful to Basil George who was chief education officer when the idea for the research was first presented, to the education committee at that time and successive committees since then, and to Pamela Lawrence the current chief education officer. Additionally, we acknowledge the help and support given by administrative and other colleagues—particularly Pat Duncan, the senior executive officer—who were working in either the Teacher Education Centre or the education offices at the Canister. Their contributions were never given without smiles and they were always undertaken without complaint. Our thanks are due as well to Tony Leo for his filming and recording work over the years. His invaluable assistance included the filming of children at play in first school playgrounds and the recording of television programs broadcast on the island. Our St. Helena colleagues on the island, particularly Cilla Thomas, Pamela Lawrence, Betty Crowie, Lilla Yon, Susan O'Bey, Evie Bailey, Betty Joshua, Corinda Essex, and Barbara George, provided invaluable assistance with data collections for the research. We acknowledge, as well, the support provided by personnel at the Cable and Wireless Office on St. Helena, particularly George Stevens (former manager) and Hensil O'Bey (the present manager). Their contributions enabled the work to continue. Without the support of our island colleagues, the project would never have commenced.

Most of all we are indebted to the staff and pupils in the island's schools who helped with the research. They gave freely of their time and energy to support us. Our regret is that we had so little to offer in return. In addition to their help and support, they left us with an insight into a remarkable community with so much to commend it. As a small token of our gratitude to them, we dedicate this book to the staff, pupils, and parents linked to the first and middle schools and the Prince Andrew School on St. Helena. We hope this volume will provide more than a cursory indication of the enviable qualities of the teaching staff, nonteaching staff, and pupils as well as the community in which they live, work, and study.

—Tony Charlton
Barrie Gunter
Andrew Hannan
December, 2000

Contributors

Daniel Charlton is Head of Communications at the South London and Maudsley NHS Trust and the Institute of Psychiatry. He graduated with a degree in Media Studies from the University of Westminster.

Tony Charlton, a psychologist, is Professor of Behavior Studies at Cheltenham and Gloucester College of Higher Education and is Director of the St. Helena TV Project. Most of his books and research articles have focused on children with learning and behavior difficulties. Since his initial visit to St. Helena in 1992, where he learned that broadcast TV's availability was imminent, he has developed a close interest in media effects on children's behavior.

David Coles is a mathematician in the Faculty of Education and Social Sciences at Cheltenham and Gloucester College of Higher Education. He is completing his PhD in the area of student teachers' mathematical experiences. With Tony Charlton, he has cowritten several research papers on the St. Helena project.

Ronald Davie is Visiting Professor of Child Psychology at Cheltenham and Gloucester College of Higher Education, a member of the Social Educational Needs Tribunal, and adviser to the Bishop of Carlisle on child protection matters. Previous positions include Director of the National Children's Bureau, Professional Adviser to the All-Party Parliament Group on Children, and Professor of Educational Psychology in Cardiff.

Barrie Gunter is Professor of Journalism and Director of Research in the Department of Journalism studies at the University of Sheffield. He is a chartered psychologist who has specialized in media research and has been published widely and has written numerous technical reports on media, marketing, and management topics. Publications include three books about children and television.

Andrew Hannan is Professor of Education and Research Coordinator in the Faculty of Arts and Education at the University of Plymouth. He holds a degree in Social and Political Science from Cambridge University and a PhD from the University of Leicester. His research interests include innovation in higher education.

Pamela Lawrence obtained her BEd degree at Gloucester College of Higher Education and Worcester College of Higher Education, in 1982 before returning to St. Helena to teach in middle and secondary schools. In 1991, she published a book on Island cookery recipes; a second edition of the book was published in 1997. In 1994, she was promoted to Education Officer for Middle Schools before being appointed in 1999 to Chief Education Officer in St. Helena.

Charlie Panting obtained a degree in Sociology and Social Policy at the University of Plymouth. After conducting voluntary work in an isolated community in the Dominican Republic, she was appointed research assistant, and then research officer to the St. Helena Research (ESRC funded) Project. She was also an interviewer for the National Centre for Social Research, an institute that conducts social research on behalf of numerous organizations and UK government departments.

Alexander Hugo Schulenburg is an historian and social anthropologist working on the St. Helena project at Cheltenham and Gloucester College of Higher Education. His research interests include the foundation and constitution of colonial and post-colonial societies and identities, and he is well known for his work on St. Helena, for which he received a doctoral degree. He is the author of several articles about St. Helena.

Cilla Thomas is currently an Education Officer for Teacher Training, Special Educational Needs, and Resources at the Teacher Education Centre in St. Helena. Her teaching experience includes general primary, mathematics coordinator, Special Needs coordinator, and Emotional and Behavioral Difficulties. During the past several years she has become increasingly involved in the St. Helena project both as a field worker, researcher, and writer. In 1999 she completed her MEd at the University of Bristol.

CHAPTER ONE

Background of the Research Project

Tony Charlton
Cheltenham and Gloucester College of Higher Education

Barrie Gunter
University of Sheffield

This chapter focuses on two distinct yet connected themes. First, the chapter briefly reflects on island history to help illustrate the sociocultural context within which the St. Helena Research Project is grounded. Second, the research project itself is introduced and aspects of it are discussed in more detail before the various substudies are highlighted, many of which are inextricably intertwined.

AN "OASIS" IN THE SOUTH ATLANTIC OCEAN

In the middle of the South Atlantic Ocean occupying a landmass of only 122 sq km, nearly 2000 km from the western shores of Africa, and almost 3000 km from the east coast of South America, lies one of the world's most isolated, inhabited islands: St. Helena. Although its location is set inside the tropic of Capricorn, the surrounding sea is somewhat cooler than expected with the water temperature averaging 19.5–21.5° by the end of winter and 24.5–25° in the summer. In his treatise on the fish and fisheries of the island, Edwards (1990) talks of the diversity of marine life inhabiting these cool waters including whales, turtles, devil rays, conger eels, sharks, and dolphins. Elsewhere, in the preface of his book, he provides an outline of St. Helena and reflects on an island that:

Erupted a few hundred kilometres east of the Mid-Atlantic Ridge as a great
volcano about 15 million years ago. . . . Above water the craggy volcanic
cliffs, often rising a few hundred metres sheer from the sea, create a spectac-
ular coastline. Only 17 kms long by 10 kms wide the island rises to 818
metres above sea level giving rise to stunning scenery. (p. 1)

Despite the diminutive dimensions of the land, its undulations add to its
size; steep hills and deep guts (valleys) amplify its ordnance survey measurements.

Notwithstanding its genesis many million of years ago, it is only in recent times that the island has experienced human occupancy of any per-
manence. However, in spite of this recency, St. Helena Island's past remains both expansive and engaging: a past that has been chronicled in a
surprisingly large number of books. Gosse (1938), for instance, provides a
compelling account of the island's history, as does Cross (1980), whereas
Evans (1994) gives an intriguing and well-documented record of education on St. Helena up until the mid-1990s. Those with ecclesiastical interests may well be attracted to Cannan's (1991) book on churches in the
South Atlantic, and others more curious about marine life can turn to Edwards's (1990) illustrated work about fish and fisheries.

Much of St. Helena's more distant past must have been forlorn, impov-
erished in fauna and flora, and devoid of animal life (marine and otherwise). Nonetheless, several million years later from around the time of the
16th century onward, the island gradually gained prominence as it became an important port of call for the sailing ships traveling from the East
Indies. For the crews of boats at anchor in James' Bay, the island supplied
rest, fresh water, and when available, fruit and fresh meat. Opportunities
also existed for harvesting the sea and "a few fish were often caught" (Edwards, 1990, p. 4). Yet, the frequency of the sailing ships' visits did little to
help establish a residential community of any size on the island. Apart
from the occasional sick mariner and escaped slave, residents of any per-
manence remained infrequent.

St. Helena's period of austere independence ended in 1633 when the
island was seized by the Dutch. Yet only a short time after this seizure, St.
Helena was claimed by the English East India Company whose representa-
tives (it is believed) were given permission by the Lord Protector Oliver
Cromwell to colonize and fortify the island. This took place in 1659. It was
nearly 2 centuries later before St. Helena became a Crown (UK) Colony,
and today St. Helena together with its two dependencies, Ascension Island and Tristan da Cunha, form a single territorial group under British
sovereignty known as "St. Helena and its Dependencies." A UK appointed
governor administers the government of the island, with assistance from
the island's legislative and executive councils; both dependencies come
within the jurisdiction of the St. Helena's governor through the offices of a

locally based administrator. Since the seizure of the island by the Dutch, a series of events have abetted the evolvement of today's indigenous population, commonly referred to as "Saints."

SAINTS

The ancestors of today's Saints were drawn to the island from disparate parts of the globe and represent a kaleidoscope of cultural and religious backgrounds. In earlier days the East India Company settled soldiers and farmers on the island. Many of the planters or farmers came to St. Helena accompanied by their wives and children, and some either brought slaves with them or purchased them later. Over the years, settlers of this kind intermingled with others who disembarked (voluntarily or otherwise) on the island. For example, efforts by the UK government to break up the slave trade on the west coast of Africa through the seizure of ships engaged in such trade brought an unexpected influx of people. Vessels seized in this way were brought to St. Helena and resulted in over 10,000 liberated slaves being shipped ashore where they were given shelter and food until they were ready to be shipped out to the British West Indies where employment awaited them. Some of the slaves elected to remain behind on the island to work as house servants and laborers. Later other more notable "guests" were brought to the island, albeit on this occasion, against their will. This natural island fortress, insulated from other continents by the immensity of the South Atlantic Ocean, provided a natural high-security penitentiary for distinguished individuals and others whom the government wished to incarcerate (chiefly to remove them from positions of influence that were considered unacceptable to the English government). After his defeat at Waterloo in 1815, Napoleon Bonaparte—escorted by a complement of advisory and other staff—was exiled to the island. Many of the island's fortifications were added around that time to help fend off those who might attempt to end Napoleon's incarceration, return him to France, and reestablish his former power, authority, and policies. Gosse (1938) indicates the magnitude of this defensive "wall," which included fortification by troops.

> First of all, with the squadron accompanying HMS *Northumberland* came the 53rd Foot Regiment. Later, the 66th Regiment from India arrived and still later, in 1819, the 20th Foot Regiment. Beside these British regiments there were the H.E.I.C. St. Helena artillery and infantry regiments. (p. 270)

Even so, it was not a triumphant rescue that finally snatched Napoleon from St. Helena but death. He died in 1821 at Longwood and was buried in Sane Valley on the island. In 1840, Napoleon's remains were returned to France. Some years later in 1890, one of Cetewayo's sons—the Zulu chief, Dinizulu—was imprisoned with his wives at St. Helena. Shortly afterward in 1902, around 6,000 Boer prisoners of war were held captive in tented encampments on Deadwood Plain and Broadbottom; the Boer senior officer General Cronje was among them. By 1903, many of the Boers—but not all, for some elected to remain behind—had returned to South Africa. On earlier occasions, there was the "import" of indentured Chinese laborers in 1810 and the practice of "charging" ships to Madagascar each time they anchored in the harbor at Jamestown. The charge demanded a slave as payment for each anchored ship.

Each of these sojourns and others on the island (whether enforced or otherwise) usually resulted in some settlers, their families and servants, mariners, soldiers, prisoners, and slaves remaining on St. Helena. Thus, the early foundations were laid for the constitution of an indigenous population. One interesting impression of this ethnic and cultural mix is given by Cross (1980) when he traces the origins of islanders' names. He talks of clearly English surnames such as Green, Young, Williams, and Reynolds before referring to names that originated when slaves were given names by their slave owners and were in:

> . . . many cases known by a single classically derived appellation and when this has been kept as a surname the lineage is apparent, eg Scipio, Mercury, Caesar, Leo, Augustus, Plato, Constantine, Joshua and Isaac. To find Caesar and Plato playing in the same cricket team has caused amusement to more than one scholarly visitor. (p. 90)

Furthermore, a plenitude of evidence is located in the *Cable & Wireless Telephone Directory* of islanders' surnames that derived from the first names of the landowners for whom their foreparents once worked (e.g., Thomas, Duncan, Henry, and Benjamin).

Given the nature of the influx of people from different parts of the globe to this small island often overlooked in such a vast ocean, it is unsurprising that today's Saints number among their foreparents a multitude of ethnic origins. Circumstances have encouraged this remarkable mix to blend into an indigenous population that, for the most part, has fashioned a coexistence that allows individuals to live and work alongside each other with a degree of equanimity and a kind of good will that is uncommon in most other parts of the world (see chaps. 2 and 6).

EDUCATION ON ST. HELENA

Arrangements for education on St. Helena are well organized and often reflect policies and practices within UK education. Between 1947 and 1987, schooling was arranged around a two-tier selective system with transfer to secondary education for 11-year-olds. High academic performers were given opportunities to proceed to the secondary selective school where they studied subjects to O and A level. Additionally, in 1973 nursery education became available from the age of 3½. In 1988 the two-tier system was superseded by a three-tier system composed of first schools (age 5–8), middle schools (age 9–12), and a single nonselective secondary school (age 13+) for a pupil population that averaged around 1,100 pupils between 1992 and 1996 (see Evans, 1994, for a history of schooling in St. Helena). In 1998 the admission age to nursery classes was lowered to 3 years (albeit entry was permitted only at the beginning of the academic year). All nursery pupils admitted must have their fourth birthday during the academic year. Although attendance at nursery classes is not obligatory, the overwhelming majority of youngsters take advantage of this facility.

In 1990 when the TV research project was first beginning to take shape, there were 7 first schools, 3 middle schools, and a secondary school. The first schools were sited close to the more populous communities. The smaller communities of Blue Hill, Levelwood, St. Pauls, Sandy Bay, and the larger communities of Longwood, Half Tree Hollow, and Jamestown (the capital) had their own first school with nursery provision. Over the next few years, decreasing school rolls and a trimming of Education Department expenditure forced the closure of the Blue Hill, Sandy Bay, and then Levelwood first schools with the pupils being bused to other first schools. Middle schools remained in close proximity to the most populous areas of the island at Longwood, Jamestown, and St. Paul's; the island's only secondary school, the Prince Andrew School, is built on Francis Plain and occupies one of the few flat areas on the island. Pupils leave the middle schools at age 12+ to transfer to the secondary school, which has a technical element attached to it. At the time of writing, plans are afoot to again reorganize the educational system; this time changing the age of transfer in all sectors from September 2000. First schools will cater for pupils from age 4–7, middle schools from age 7–11, and secondary school pupils will range from age 11–18. Plans are also in hand to raise the school learning age to 16 in the year 2001. On this occasion however (unlike practice in the earlier two-tier system), secondary education will remain nonselective.

All of the central organization for the educational system is undertaken in the Canister at the top of Main Street—the "home" of the chief education officer, administrators, and most of the other education officers. Further up from James' Bay, in the direction of Ladder Hill, is the Teacher Education Center where teacher-trainees follow a 2 year course with a further 2 year probation period in schools. Nursery assistants undertake a 1 year training course followed by 1 year probation.

THE RESEARCH PROJECT

The genesis of the St. Helena Research Project was a fortuitous one. In 1990, one of the editors (Tony Charlton) was working on St. Helena for a 6-week period. On that occasion and again in 1992, he was employed as a consultant (i.e., as a psychologist) by the UK Overseas Development Administration, now called the Department For International Development (DFID). His brief was to assist the island's teachers with the identification of those children who were evidencing special educational needs in terms of their emotional or behavioral functioning and in making assessments of, and then devising associated programs to help meet, identified learning needs. On the second visit, in 1992, he discovered that plans were afoot to introduce broadcast TV services to the island, although it was unclear at what time these services would be introduced. The opportunity to undertake a pre- and post-TV research project of this kind in an almost unique setting was compelling and one that was readily grasped. Arrangements needed to be devised that would permit a study of a limited kind to proceed in the first instance (given the lack of financial resources available at that time), but allow opportunities to expand the research at a later time. Taking into account the island's isolation as well as the restricted opportunities to travel to and from the island, it became clear that any plans to undertake a research project of this kind needed to be organized and dispatched with some haste: timing was essential because a follow up visit was unlikely within the next 12 months. Moreover, any arrangements had to be undertaken in a manner that did not adversely impinge upon the consultancy responsibilities.

As part of his work in the island's schools, Charlton had already obtained teachers' ratings on the behavior of first and middle school pupils, which indicated that the young children were uncommonly well-behaved. Some of this information was used later as part of the pre-TV data collections. The consultancy duties at that time made it difficult for him to become heavily involved in collecting other data. But with the help of staff in the secondary school, and the first and middle schools in particular, as well as colleagues from the Education Department and from elsewhere on the island, he was able to begin making arrangements for other additional data collections. Although teachers' ratings were used to obtain information on pupils' behavior, an over-reliance on this type of measurement was ill advised. Other forms of measurement needed to be incorporated, including observations of children's behavior in the playground and in classrooms (measurements of a kind which Scott, 1996 refers to as "Gold Standard" measures). In the short time available, it was neither practical nor practicable to train observers to collect data on children's antisocial and prosocial behavior in playgrounds. Subsequently, a compromise was reached and arrangements were made for the video recording of free-play behavior in playgrounds of two of the larger first schools in Half Tree Hollow and Jamestown. (The behaviors on these early videotapes remained uncoded and were not analyzed until 1997. Later video recordings tended to be coded as soon as they arrived in the UK.) A decision was also made to obtain additional pre-TV data on pupil behavior in classrooms, on nursery class pupils' video-viewing habits, about middle school pupils' leisure-time pursuits, and on first and middle school teachers' perceptions of problem behavior in classrooms and around school.

The observations in first and middle school classrooms that focused on 6- to 10-year-olds were difficult to undertake because experienced observers were not available. However, with the agreement of the chief education officer, 8 first and middle school teachers were trained to use Merrett and Wheldall's (1986) Observing Pupils and Teachers in Classrooms (OPTIC) coding schedule, so information was collected not only on young children's on-task behavior in classrooms but also on selected aspects of class teachers' behavior. Observers collected data only in schools other than their own. Elsewhere, in order to obtain information on nursery class children's video-viewing habits, preschoolers completed video-viewing diaries showing how much time they viewed as well as the type of programs they watched. In first and middle schools, teachers completed questionnaires asking for their perceptions on pupils' around school behavior (e.g., how much and what kinds of misbehavior did the teachers encounter). Finally, in order to investigate not only content effects but also displacement effects, information was collected through questionnaires about children's (age 10–12) leisure-time pursuits, where they undertook them, and with whom. Information from this source also provided data on the leisure-time activities they liked best.

In the pre-TV phase, the following data were gathered for later comparison with post-TV data (collected at several points in time between 1995 and 2000):

1. Teachers' ratings of nursery class, first, and middle school children's behavior (checklist completion for each pupil);
2. Teachers' perceptions of first and middle school pupils' behavior (questionnaire completion by each teacher);
3. Middle school pupils' leisure-time pursuits (questionnaire completion by each pupil);
4. First school pupils' free-play behavior in playgrounds (videotape recordings for later coding and analyses of group level data);
5. First and middle school pupils' on-task levels (observations in classrooms on small groups);

6. First and middle school teachers' classroom behavior (observations of individual teachers); and

7. Nursery class pupils' video-viewing habits (diary completion by each pupil).

In the post-TV phase from March 31, 1995 to roughly 2000, the above measures (apart from 2, 5, and 6) were readministered and augmented by:

1. Content analyses of television programming (recordings of all programs over 6 days in 1997 and 1998);

2. Focus group discussions with 16- and 17-year-old students at the Prince Andrew School (audiotape recordings); and

3. Pupils' records of their broadcast television viewing habits (diary completion by 8-year-olds).

All data were analyzed in the UK.

RESEARCH PROBLEMS

From its conception in 1992, the research project encountered a number of problems most of which were logistical. Some of these were relatively easy to overcome; others were more troublesome. The source of the most pressing problem was the remoteness of an island located some 2,000 km from the western coastline of the African continent. Inextricably linked to this isolation were transport difficulties. Although air transport can take one most of the way to the island (i.e., to Ascension or Capetown), the remainder of the journey is dependent wholly on one ship, the Royal Mail Ship *St. Helena*, a luxury cargo vessel with cabin space for 128 passengers. The RMS travels regularly between Cardiff (UK) and Capetown (South Africa), calling at Tenerife, Ascension, and St. Helena en route. When the ship leaves St. Helena on its voyage north to Cardiff, it is normally around 6 weeks before mercantile links are restored to the island (e.g., unloading mail, groceries, clothes, machinery, furniture, fabrics, and passengers). The longest route involves a 2-week journey from Cardiff sailing on the RMS to the island by way of Tenerife and Ascension Island; the quickest route (usually, but not always) is by air to Capetown and then embarking on the RMS for a 5-day passage north. An alternative route requires boarding the Falklands' flight from RAF Brize Norton (UK) but landing at Ascension Island and remaining there for a few days before embarking on the RMS for the 2-day voyage south

In reality, none of these itineraries was without difficulties. At times, the RAF flight was delayed by inclement weather or mechanical failure somewhere along its flightpath to and from Mount Pleasant in the Falklands. Outbreaks of hostilities in the Middle East and elsewhere meant that flights were often fully booked with troops returning from Mount Pleasant, and the RMS occasionally broke down. (On one occasion, Tony Charlton arrived at the airport at Cape Town and was told by the taxi driver who was taking him to the city that the RMS was in dock with a hole in the bottom! Thankfully this tale proved to be untrue, although the ship's departure was held up because of rudder problems.) Predicaments such as these could thwart the best arrangements and delays could have grim consequences. Travel obstacles of this sort usually reduced the number of project "working" days on St. Helena; they could also impose additional expenditure on a modest budget and present frustrations.

One other constant problem concerned publicity. In unison, the island's remoteness, its rich heritage over the last 4 centuries, and its uncommon delay in receiving broadcast television combined to make the research attractive to newspapers, radio, and television. On those occasions when media attention was drawn to the study, the island's remoteness presented them with substantial difficulties. More often than not, the island's relative inaccessibility dissuaded the media from traveling to St. Helena. One agreeable consequence of this hindrance was that reporters often had to rely upon the researchers to name islanders whom they could interview. This arrangement suited the team in their efforts to manage the media in ways that helped forestall inaccurate and misleading coverage of the study. This strategy was assisted by carefully planned press releases that were prepared in detail and usually accompanied a questions and answers (Q & A) information sheet to keep reporters well-informed on research results. (The publicity machinery and press coverage is considered in more detail in chap. 7.) In consequence, much of the research was reported in the press without interviewing the research team. Alternatively, reporters made use of the press releases, the Q & As, as well as radio interviews with islanders recommended to them by the researchers. Of the hundreds of press reports, fewer than a handful produced inaccurate or misleading coverage.

On other occasions, the time required to travel to and from St. Helena (normally at least 6 days), as well as the travel and subsistence costs associated with these visits, restricted the number of trips that were undertaken by the UK-based research team; however, several trips did take place between 1990 and 1998. This meant that much of the research data in the post-TV phase had to be collected by someone other than project team members. Fortunately, the team had developed much valued links with the Education Department and especially with the island's schools. Education officers and schoolteachers gave generously of their time and energy to assist with the research. In particular, help was much valued from Lilla Yon (now education officer, primary schools), Cilla Thomas (now education officer, training), Betty Crowie, Betty Joshua (now head teacher of Two Boats School on Ascension Island), and Susan O'Bey (now education officer, head of Prince Andrew School). All five have been educated to masters degree level and have successfully completed a two-term program in Methods of Educational Inquiry. Hence, arrangements were in hand for some data collections to take place in the absence of a UK researcher, although all data were always coded, entered, and analyzed in the UK. Other key assistance was given by Tony Leo (the manager of the radio station in St. Helena). He arranged both for the video recordings to take place in the 2 first school playgrounds and the recording of broadcast television programs so that they could be analyzed for violence and prosocial content in the UK. Without this help, it would have been difficult (if not impossible) for key aspects of the research to be undertaken.

Another enduring concern was linked to funding. The study was initially supported in part by faculty research moneys from the UK academic base, Cheltenham and Gloucester College of Higher Education and the University of Plymouth (through Andy Hannan). Other moneys were added from the proceeds from articles written for newspapers and radio interviews. However in late 1994, discussions with *The Sunday Times* helped to secure some financial backing for the research from 1995 to 1997 when a grant was obtained from the UK's Economic and Social Research Council. This funding allowed the study to proceed on sound financial footing for the next 3 years and facilitated the appointment of a research assistant for the final 2-year period. The research team had already been strengthened in late 1992 by the addition of Andy Hannan (to work on the displacement study), and Barrie Gunter joined the project in 1995 (to focus primarily on the content analysis of TV programming). Ron Davie was welcomed into the team in 1996. Later, in 1998, Charlie Panting was appointed as the research assistant (later changed to research officer). With this last appointment, the research team was complete. It included four principal researchers in the UK (although there were others also involved who worked on the project from time to time) together with colleagues from the Education Department on the island.

PRIOR EXPOSURE TO VISUAL MEDIA

Over the years, it has become an increasingly rare event to encounter a "westernized" community that has not experienced regular access to visual media of one kind or another (e.g., cinema, video, or television).

Most communities have been exposed to all of these media (for a lengthy period of time in most cases). Furthermore, for the large majority of communities there is a common logical sequence in their access to the various media. In most instances, the cinema appeared first, much later followed by the television, and then by the video.

Since television's availability in the 1950s, much improved technologies linked to both the transmission and reception of broadcast signals have helped popularize television. Improvements of this kind particularly in the 1960s and 1970s, so enhanced TV reception opportunities that those westernized regions and communities, for example, hitherto denied access to TV (mainly because of their geographical location or isolation), were able to access it at last. As a consequence, within the last few decades it has become more and more difficult to locate a (broadcast) TV-naive community in the westernized world. Hence, most naturalistic studies of TV effects have had to make use of communities that—although not TV-naïve—have had only limited access to television. For example, William's (1986) North American study focused on a small community (given the nom de plume "Notel"), deemed not to have access to TV but soon to have access to it. Prior to the availability of "faultless" television reception, the town was located in a geographic blind spot that hindered full-strength TV transmitter signals reaching the town. Yet in reality, at least some households were picking up broadcast TV signals, albeit weak ones. Additionally, some of the inhabitants were viewing as guests in houses other than their own either within Notel or elsewhere. The town had also enjoyed long experiences with the cinema. Even so, despite the confounding intrusion of viewing experiences of this kind (in research terms), access to broadcast television was a novel encounter for most of the Notel population.

In St. Helena, for economic considerations together with technological reasons linked to the island's remoteness, the natural progression in the availability of the various media was transposed. The modest size of the island population offered few, if any, commercial incentives for others to provide costly satellite arrangements for television reception. Nevertheless, although the video preceded the arrival of broadcast television, other visual media arrived in a logical and expected order. Magic lantern presentations were available and popular from the 1900s onward, and the first cinematograph show was given in 1927. Later in that year, large cinema audiences were attracted by the on-screen appearances of Buffalo Bill Jr. and Buddy Roosevelt. Even at that time, however, exposure to cinema generated anxieties among some islanders about untoward outcomes of the screen's influence. *The St. Helena Diocesan Magazine* (1929) wrote of such shows as being "more baneful than beneficial" (p. 152). Whether baneful or beneficial, for some in the audience there was an added incentive to sitting before the "large" screen. A glimpse of the grandeur, opulence, and "spirit" of the "accommodation" (reserved for a few chosen ones) on those viewing occasions is captured in the same journal with the following report:

> We sat upstairs in armchairs with cushions, on one side of the gallery, whilst the Governor with his party sat on the other. They were all in evening dress! The cinema did not start until the Governor had arrived, which was about 8-30 p.m. . . . . Then the St. Helena Band started, a stringband consisting of three players. They had no sooner started to play than the whole audience took up the tune and sang lustily. We'll never forget, "I'm Sitting on Top of the World," or "I'll be Loving You Always," or "Show Me the Way to Go Home." (p. 61)

In 1940 the "talkies" arrived. The first film was entitled *Shipmates Forever* and featured Dick Powell. An indication of the popularity of the cinema at that time is shown by the several cinemas available to a population of only around 5,000: the Ladder Hill Drive In overlooking Jamestown and the Queen Mary Theatre in Napoleon Street in Jamestown (which could house over 500). Sadly with the advent of the video, the public screen became obsolete. All the cinemas had closed by the mid-1980s.

The video first arrived in 1979. Although video hire shops were soon established in Jamestown and surrounding communities, videotapes were also sent to the island from the UK and South Africa. According to the island's 1987 census, 29% of households owned one or more video sets. Although financial circumstances could limit the widespread availability of the sets, so could the unavailability of electricity. At that time many households, particularly those located in more remote and less accessible regions on the island, were not connected to a main electricity supply. Even so, in 1994 (the year before broadcast television's availability) most 3- to 4-year-old children were watching videos for around 11 min per day and much of their viewing diet consisted of cartoons (e.g., "Tom and Jerry" and "Woody Woodpecker").

THE LAUNCHING OF THE TELEVISION
SERVICE ON ST. HELENA

On March 24, 1995 the *St. Helena News* announced that at 9:00 a.m. on Friday, March 31, 1995 the St. Helena Television Service would be launched. The island's governor would open the service by talking to islanders on the television. The weekly paper carried the following account of the television service to be provided:

The base station has initial capacity for two channels but the service will start with only one in use. The 24 hour continuous news and information channel, CNN International Program, will be available from 31st March with the second channel coming into service in September 1995, following the launch of a new satellite. It is planned that this second channel will carry programmes from the BBC as well as other entertainment channels. (p. 1)

By April 1996, the *St. Helena News* talked of a subscriber base of around a third of householders; additionally, a large number were receiving the service but not yet paying for it. On November 1, 1996, an expanded service was made available including the additional services of Cartoon Network, Hallmark (films), Supersport, and Discovery. One more change was made in June 1998 when the program line up was changed to M-Net Brochure (KTV, Movie Magic, and BBC), Discovery, and Supersport. The addition of more transposer sites meant that an increasing number of islanders had opportunities to tune in. In March 1998, the service became fully encrypted; from this point onward reception required a decoder.

BROADCAST TELEVISION VERSUS THE VIDEO

Although the St. Helena TV Research Project monitors children's social behavior across the availability of broadcast television, it is important to consider how islanders' prior experiences of cinema and video may have intruded on the current study. In one very fundamental yet central sense, they did not. The aims and objectives of the St. Helena TV Research Project were specific; the focus of the project was restricted to a search for changes in social behavior across the availability of broadcast television. Given ubiquitous claims that viewing videos encourages some viewers to act in violent and other antisocial ways (see Gunter, 1998, for review), some may question the value of undertaking the St. Helena research in the sense that—according to video critics—video viewers on the island were already likely to have been corrupted by their viewing experiences. There are at least four plausible responses to such a query.

First in the pre-TV phase, nursery class children (i.e., 3- and 4-year-olds) and middle school pupils (i.e., 9- to 12-year-olds) were found to be among the best behaved worldwide (Charlton, Abrahams, & Jones, 1995; Charlton, Bloomfield, & Hannan, 1993). Furthermore, 7- to 10-year-old pupils were found to spend more of their class time on-task (doing what the teacher had instructed them to do) than most of their overseas peers (Charlton, Lovemore, Essex, & Crowie, 1995). Moreover in the first and middle schools, teachers' perceptions of the kinds and extent of pupil misbehavior they encountered suggested that such problems were less frequent and less serious than those experienced by their overseas colleagues (Jones, Charlton, & Wilkins, 1995). Given this commendable behavior by pupils in the pre-TV phase (even though many of them had access to video), it is difficult to consider a more testing baseline against which to search for any pernicious effects linked to live television viewing.

Second in the pre-TV phase, nursery class children (who were a central focus in the study) were found on average to be viewing videos for around only 11 min daily and most of their viewing diet was confined to cartoons. (It is worth noting that in 1987 although around 30% of households had videocassette players and recorders, the recorder function was redundant because of the absence of live programs to record. The recording and associated playback facility became functional only with the arrival of broadcast television in 1995.) So, neither the amount of time given by these children to their video viewing or the nature of their viewing diet was of a kind indicating any serious cause for concern according to findings in much of the voluminous literature on adverse viewing effects for video or TV.

Third, by comparison with seeing videos, watching television (particularly where more than one channel is available) can encourage more uninterrupted viewing. Individuals can persist in viewing television even when their preferred program finishes. This involuntary viewing can be tempting because it requires minimal effort other than "attending." Individuals can acquire the "habit of viewing out of ritual to fill time or to relieve boredom" (Rubin & Bantz, 1987, p. 483). Along these lines Dorr and Kunkel (1990) remark that the "long-standing axiom of television-viewing behavior, for both children and adults has been the decision to watch typically supersedes the decision of what to watch" (p. 7). This easeful and effortless habit is less easy to sustain with the video for the operations entailed can make the varied tasks wearisome and laborious. With video viewing, labor is usually required, for example, for selecting and renting a video; the physical loading into the player; operating the video controls; changing and storing cassettes at the end; and returning the video if rented. Tasks such as these can help discourage immoderate amounts of video viewing (particularly when rental expenses are considered). Thus, the television is the more likely of the two services to encourage persistent or even heavy viewing. The provision of more than one channel and service (e.g., CNN, BBC, and M-Net) makes persistence viewing more likely. What is more, the television effects' literature suggests the amount of viewing (particularly where this is heavy and includes violent programming) is an important predictor of harmful effects occurring (Abelman, 1991; Singer & Singer, 1986a).

In summary, the above remarks support the merits of investigating viewing effects across the availability of broadcast television in St. Helena for several reasons. First in the pre-TV phase, young children in St. Helena tended to be only light viewers of videos (mainly cartoons). Second, young children's good behavior before the availability of television gives little, if any, indication that they had been affected adversely by their video viewing experiences. Finally compared to the videoplayer, television is more likely to encourage viewing for longer periods of time (e.g., because of its facility to "kill time" and encourage persistence viewing).

Even after their limited experiences of watching videos, in the pre-TV phase of the study young children in St. Helena remained uncommonly well-behaved and seemingly outwardly unaffected by their video viewing. However, the availability of broadcast TV threatened the maintenance of this good behavior because it can encourage viewing additional to viewers' preferred choice; the more they watch, the less selective they become (Sun, 1989). Consequently, this involuntary viewing increases risks of adverse effects stemming from both extended viewing and heightened exposure to antisocial TV (see chapter 6). Hence, this "good behavior base" in the pre-TV phase provided a useful baseline against which to test for viewing effects across the availability of broadcast television. Likewise given that the availability of broadcast television was likely to increase viewing times, it seemed appropriate to study how time was displaced from other leisure-time pursuits and dedicated to viewing television (see chapter 4).

THE STUDY

In a naturalistic setting, the St. Helena Research Project has monitored and studied children's behavior across the availability of broadcast television as well as across an expansion of television services (1991–2000). Both qualitative and quantitative methodologies have been used to investigate content and displacement effects linked to television viewing (although to date prominence has been given to quantitative data collections, imminent research will rely more heavily upon qualitative data). The project is comprised of a number of interlocking substudies that have all been concerned with investigating effects of broadcast television on young children's school behavior and leisure-time pursuits.

The study is remarkable in many ways:

- It has the methodological advantages of a longitudinal (quasi-) experiment being undertaken in a naturalistic setting;
- It offers exceptional opportunities in an English-speaking westernized culture to investigate both TV content and displacement effects upon (broadcast) TV-naive participants;
- The island's isolation reduces opportunities for contamination effects (e.g., pupils as "guest viewers" overseas and participant attrition); and

- Finally, the study offers scope to investigate TV's influence in a remote community to determine if pupils' isolation makes them more dependent on television or less reliant because they have become adept at devising their own pastimes (Himmelweit, Oppenheim, & Vince, 1958).

RESEARCH DESIGN

The inquiry—a longitudinal field design in which phenomena under study are observed, measured, and evaluated in a natural setting—is concerned with investigating any effects not only across the availability of broadcast television but also across an expansion of television services (e.g., CNN, M-Net, and BBC). To investigate *content effects*, we used a longitudinal *cohort time-interrupted series model* combined with a *cohort time-lag cross-sectional model* as the principal design. Although all 3- to 12-year-old pupils (and some older students) were participants in one or more of the substudies, successive cohorts of nursery class children were a major focus for the inquiry. *Displacement effects* were evaluated through a cross-sectional sequential design using 9- to 12-year-olds in 1994, 1995, 1997, and 1998. Coding and data entries from leisure-time diaries made use of the categories set out originally by Himmelweit et al. (1958).

Procedure

Most pre- and post-TV data collections took place annually in October and November. Data collections on the island were either supervised or overseen by the project researchers (Tony Charlton, Barrie Gunter, and Andy Hannan) with the help of education officers and teaching staff from the Education Department on St. Helena. A local technician undertook most of the video filming in playgrounds and made the arrangements for the recording of television programs.

Participants

In selecting participants, efforts were made not to inundate particular classes and teachers as either data collectors or data collection sources, although all pupils attending first and middle schools (ages 5–12) and some older students from the Prince Andrew School were involved in aspects of the research. Additionally, all nursery class students (ages 3–4) from 1993–1999 were followed for the life of the investigation.

Measuring Instruments

Quantitative and qualitative data were obtained using:

The Preschool Behaviour Checklist (PBCL; McGuire & Richman, 1986)

Teachers rated young children's behavior on 22 items (e.g., concentration, fighting, teasing, temper tantrums, destructiveness, and interference with others). In this study, the PBCL was used annually with preschoolers (1993–1999).

The Rutter Behaviour Questionnaire (RBQ; Rutter, 1967)

In 1992 and 1998, teachers rated their pupils' overt behavior on 26 items (e.g., truants, bullies, poor concentration, worries, fights, and restlessness). The RBQ provides antisocial, neurotic, and undifferentiated subscores. The RBQ was used with 6- to 12-year-olds.

OPTIC (Merrett & Wheldall, 1986)

OPTIC allows observers both to estimate pupils' on-task behavior and to obtain information on teachers' behavior in response to pupils' academic and social behavior. In each 30 min observation session for five periods of 3 min (Section A of the schedule), the measurement tool samples the teacher's use of approvals and disapprovals to pupils' academic and social behaviors. In the remaining 15 min (Section B) observers sample pupils' on-task behavior for five periods of 3 min. OPTIC has been validated against more lengthy observation schedules (interobserver agreement averages over 90%; Merrett & Wheldall, 1986). During 1994, it was used in the present study with 7- to 10-year-olds ($N = 169$) from three of the larger first schools and the three middle schools.

TV-Viewing of Aggressive Behavior

The content analysis study focused on depictions of violence and prosocial behavior on screen. In both 1997 and 1998, the television program output was sampled over 3 days (6 days in total). Viewing data were collected on programs broadcast between the hours of 6:00 a.m. and midnight on Tuesdays, Thursdays, and Sundays. Programs broadcast between the school hours of 9:00 a.m. and 3:00 p.m. on weekdays were excluded. The output was recorded from the two broadcast TV channels and the television services that share these channels. The analysis of the program recordings was undertaken by trained coders in the UK. The coding frame comprised normative definitions of violence and prosocial behavior and definitions of the unit analysis (e.g., violent incidents or events); attribute codes were used to identify contextual features of the violence and prosocial behavior. The latter was defined through reference to audience opinion and TV effects research.

Children's TV-Viewing Logs

For each of 3 days in 1998, 8-year-old pupils completed a TV-viewing diary showing viewing hours and programs watched in addition to program and character preferences. Individuals' program viewings were given antisocial and prosocial loadings taken from the TV program analyses undertaken on the same day as the completion of the diaries (see previous paragraph). RBQ scores were obtained for the same participants at the same time. PBCL ratings were available for the same group from 4 years earlier when they attended nursery classes.

Video Recordings of First School Children's Behavior in Playground Settings

Aggregate measures of 3- to 8-year-olds' antisocial and prosocial behavior in free-play settings were obtained in October and November 1994 (pre-TV), 1995, 1996, 1997, and 1998 (post-TV). Each year, behavior was filmed in two first school playgrounds on 10 days during morning and afternoon break times. Nearly 2,000 min was filmed in this way. Recorded behaviors were coded and analyzed in the UK using the Playground Behaviour Observation Schedule (PBOS; see Charlton, Gunter, & Coles, 1998). The schedule includes 26 categories of behavior (e.g., kicking, pushing, turn taking, consoling, fantasy play, games, observing others, affection, and helping others). Playground behaviors were then coded and analyzed with genders separated and combined. Group size was considered also (i.e., 1, 2, 3–5, 6+). In the post-TV stage, imitation of television characters was also included (e.g., imitating facial expressions, gestures, dress, and acts).

Children's Leisure-Time Pursuits Diaries

In October 1994 (pre-TV), November 1995, February 1997, and October 1998 (all three post-TV) all 9- to 12-year-old pupils were asked to complete diaries under supervision in classes about their leisure-time habits before and after the arrival of broadcast TV. They were requested to document on an hour-by-hour basis *what* activities they engaged in, *with whom* they undertook such activities, *where* such activities were undertaken and *which* three of the leisure-time pursuits they enjoyed best. Subsequent coding entries made use of categories originally set out by Himmelweit et al. (1958).

Focus Group Discussion

In 1997, a small group of 16- to 18-year-olds attending the Prince Andrew School participated in a group discussion that considered effects of broadcast television on the island. Although this discussion was relatively brief and unstructured, interesting and potentially important explanations were put forward to account for some of the findings from the other studies; findings on the whole showed little change in children's behavior following the introduction of broadcast television on the island. The students' discussion threw light on potentially mitigating environmental factors.

Teachers' Perceptions of Around-School Behavior in First and Middle Schools

A questionnaire was distributed to 54 first and middle school teachers on the island to inquire whether they thought they were spending more time than they ought to on problems of order and control and to find out the most troublesome behaviors they encountered as well as how often they encountered them.

Data from two or more of the above measures were sometimes combined. For example, data were merged from the content analysis of TV programs and diary measures of children's viewing to obtain more precise measures of the quantity of antisocial and prosocial content that children have consumed. These data were then linked to measures obtained on their behavior both before and after the availability of broadcast TV. These mergers allowed investigations that were focused on individuals rather than on being pitched at the aggregate group level as with the data obtained from analyses of the video recordings of children's free-play behavior. Both individual and group data collection approaches have their strengths and limitations; these are discussed in chapter 6.

CONCLUSION

The St. Helena Research Project took advantage of a remarkable opportunity to monitor children's social behavior across the availability not only of broadcast TV but also of an expansion of TV services. It included the methodological advantages of a longitudinal (quasi-) experiment that was undertaken in a naturalistic setting. There were other benefits linked to the study, too. These included the good behavior of young children in the pre-TV phase; opportunities in an English-speaking westernized culture to investigate both TV content and displacement effects on (broadcast) TV-naive participants; and the island's isolation, which helped reduce chances for contamination effects (e.g., pupils as guest viewers elsewhere and participant attrition). The remaining chapters in this book focus on the various aspects of the study before considering ways in which the research might be extended in the future.

CHAPTER TWO

"They Are as if a Family": Community and Informal Social Controls on St. Helena

Alexander Hugo Schulenburg
Cheltenham and Gloucester College of Higher Education

Since 1992, the St. Helena Research Project (Charlton, 1998b) has been studying the impact of broadcast television in one of the most isolated Western island societies, the British Overseas Territory of St. Helena (Cross, 1980; Foreign and Commonwealth Office, 1999; Schulenburg, 1998). Research presented in this volume claims that nearly 5 years after television was introduced to St. Helena there has been no significant increase in levels of antisocial behavior among young children on the island. Instead, the high rates of good behavior among young pupils noted before television's arrival appear to have been maintained. These findings are contrary to claims that exposure to television encourages children to behave violently (Joy, Kimball, & Zabrack, 1986; also see Barker & Petley, 1997).

As is suggested in a number of publications (Charlton, 1997, 1998a, 1998b) and in later chapters of this book, adverse effects of the introduction of television on St. Helena have been negligible so far perhaps as a consequence of "a kind of uncoordinated pastoral network within the community" (Charlton, 1997, p. 59). Hence, Charlton (1998a) states, "a strong argument can be made that where families and communities are 'watchful' and caring over youngsters' and adults' behaviour, dominant (harmful) determinants of behavior exclude TV" (p. 29). In other words, "watchfulness encouraged individual and collective accountability; qualities which appeared both to underpin acceptable social behaviour in St. Helena" (Charlton, 1998b, p. 170). On St. Helena, exposure to violence on television is not currently a cause of violent and other antisocial behavior in young people because "a healthy family, school and community environment" (p. 171) are more important than television in shaping the behavior of the island's children.

The St. Helena Research Project's findings do not question claims that viewers learn antisocial behavior from television. Instead, they highlight the role of environmental cues in mediating between the learning and practicing of antisocial behaviors. According to this hypothesis, children will be less inclined to perform antisocial behaviors in settings where social controls such as watchfulness and care are exercised over them. Rather than reproaching television in cases where antisocial behaviors are evident, more often than not it is the absence of reasonable and consistently applied social controls that offer a more likely explanation (Charlton, 1998b; Laub & Lauritsen, 1998).

To substantiate these early, tentative thoughts on the nature of St. Helena society and its role in maintaining levels of good behavior, this chapter presents details of qualitative research into young people's accounts of social and community relations on St. Helena. But first, a context for this research is provided by a review of relevant debates on informal social controls and the character of small communities.

SOCIAL CONTROLS

Despite being far from novel (Downes & Rock, 1995), theories of social control or control theories come in a wide variety of guises (Brewer, Lockhart, & Rogers, 1998; Green, 1982). Although macrosocial perspectives explore formal modes of control (including government, the legal system, and law enforcement), microsocial perspectives focus on informal modes of control, such as reputation and the family. With respect to the latter, most authors agree with one version or other of Hirschi's (1969) claim that the common property of social control theories is their assumption that "delinquent acts result when an individual's bond to society is weak or broken" (p. 16).

An insightful, yet neglected variation of Hirschi's microsocial approach can be found in Tittle's work on sanctions and sanction fear as contributors to social order. Tittle (1980, p. 42) defines deviance or nonconformity as "behavior that is considered unacceptable, inappropriate, or morally wrong by the majority of people in a given social group," and suggests "community cohesion" as a principal variable in explaining conformity. Cohesion, according to Tittle, "grows out of frequent mutual interaction and collective cooperation in meeting challenges, both of which become logistically more difficult with larger numbers" (pp. 280–283). He furthermore argues that "cohesion is reflected in a sense of community spirit," two indicators of such cohesion being the "number of residents in the locale" and the amount of "community spirit."

A related approach to questions of order and cohesion in society is evident in network analysis and rational choice theory, both of which suggest that it is in the self-interest of all individuals to maintain social order. Extensive research into questions of social cohesion or "solidarity" is currently being undertaken by the Institute for the Study of Cooperative Relations (ISCORE) at the University of Utrecht. According to ISCORE (1999), "solidarity is a *pattern* of behavior within groups: helping others in *need*, contributing to *common goods*, resisting the *temptation to breach* agreements at the expense of others." This pattern of behavior arises in specific forms of medium- or long-term collective action in face-to-face groups.

Indeed, perhaps the most important feature of small communities is that their members interact in stable and durable relations. If people know for sure that they will interact with each other in the future, then they will have no incentives to cheat on others or display distrust. According to Raub and Wessie (1990), dense social networks, which are a typical feature of small communities, make a reputation of trustworthiness particularly useful. Furthermore, a common past provides individuals with learning opportunities to establish which partners can indeed be trusted. Violating social norms is costly if others observe or perceive this violation, for any such violation reduces the availability of resources possessed by others, which they would be willing to share if one could be trusted (i.e., if one had a good reputation).

Tittle (1980, p. 179) in particular has focused on what he calls "reputation-effect" as a means of social control, identifying nine specific indicators of different aspects of "the probability of . . . suffering negative consequences." These include:

> 1. The chances that somebody who does not approve of it would find out (discovery); 2. the chances that people known personally would find out (interpersonal exposure); 3. the chances that most people in the community would find out (community exposure); 4. the amount of respect the respondent would lose among people known personally if they found out about it (interpersonal respect loss); 5. the amount of respect the respondent would lose in the community if they found out about it (community respect loss).

In the words of Kiesler and Kiesler (1969), "whether a person will continue to comply when a group is not present depends partly on whether he feels that the group would be aware of deviant behavior" (p. 63).

Although Tittle (1980) concludes that "the most important component of sanction fear is that derived from interpersonal and other nonlegal sources" (pp. 180–182), that is, from informal sanctions, most recent research into "surveillance" as a mechanism of social control has focused on formal and technological methods or at least on processes in which the observer is socially distant (in location, status, or involvement) from the person being observed. Research on the importance of surveillance or monitoring for informal social control has been somewhat scarce in recent years. Yet, there are exceptions. Borg (1997, p. 273), for one, has examined "the concept of social monitoring, defined as the process of collecting, storing, and exchanging conflict information," and suggests "a framework for quantifying the process" and proposes "several factors associated with its variation." Overall, Borg concludes "that the quality and quantity of social monitoring within a group varies in response to the social distance, social status, and interdependence among its members." According to Richard W. Wilmack (personal communication, 1999), it hence "seems reasonable to hypothesize that the efficacy and efficiency of informal social control depends on the ability of other people to witness, monitor, or otherwise become aware of the behavior to be controlled." Drawing on the results of research into alcohol abuse, Wilmack (personal communication, 1999) has stressed the "inhibiting effects of multiple role obligations," which "make a person's drinking behavior and its consequences more readily observable to more people." In consequence, it can be argued that "in a small community with relatively high rates of interaction among the community members, drunkenness or any other overt and abnormal behavior would presumably be more conspicuous and more widely witnessed." Similar conclusions were drawn by Linden, Currie, and Driedger (1985) and are also evident from a longitudinal study of the role of parental monitoring and of family cohesion in preventing alcohol abuse and delinquency (Barnes, 1995; Barnes & Farrell, 1992, 1993). The importance of informal surveillance can hence be studied not only at the level of "the community," but also at the level of smaller groups such as the family. Research by Langellier and Peterson (1993), for instance, has focused on "family storytelling as a strategy of social control" (p. 49).

According to A. P. Cohen (1982b), what he terms the "publicness of knowledge" has "profound consequences for the conduct of social life. It characterizes the elaboration of culture, the organization of social structure; it pervades all the ways and circumstances in which people confront each other socially" (p. 10). The primary consequence of this publicness of knowledge is the achievement and maintenance of social conformity or discipline. Cohen argues, "because people know a great deal about each other and because everything is salient as public knowledge, people have to behave in particular ways" (p. 11).

In a study of parenting in a rural community, Valentine (1997) found that local parents explained that in "a village where everyone knows everyone else, they could feel confident that there were always 'eyes on the street' to keep their children under observation," although "some argued that the gaze of nosy neighbors could be rather too intrusive" (p. 144). A result of such nosiness is gossip, which is not only the principal medium of the flow of information underlying public knowledge but also a key mechanism of social control (Jones, 1999; McFarlane, 1977). According to Scott (1990), gossip reinforces "normative standards by invoking them and by teaching anyone who gossips what kinds of conduct are likely to be mocked or despised" (p. 143). Moreover, gossip is used "to include and exclude; . . . to judge and constrain others; to confer and deny favors" (Emmett, 1982, p. 298). Consequently, the "degree of knowledge of each other which small-town people have, the lack of privacy, can be and often is oppressive and felt as constraining."

With respect to St. Helena, the relationship between the inevitable intimacy of social relations and the resultant existence of social controls has been explored in a short story, "The Pepper Tree," by Basil George (1960), one of the island's few local authors.

> The walls of the valley rose sharply on three sides of the town, making it one big house with a sky roof and a sea door. . . . The people lived secure within the shelter of the valley like one big family. And like one big family their cares and tears and fears hung scattered on a clothes line. (p. 2)

Under these circumstances, the story's central character, a boy called Numes, "would learn his life could not be contained within himself in a neighborhood where you don't dry clothes indoors" (p. 2), that is, where it is nearly impossible to keep family matters private.

The effectiveness of social controls in small communities has indeed been found to be largely attributable to their "restrictive, bonding character" (Crow & Allan, 1994, p. 11). Yet, without such mechanisms says Elias (1974), a locality

> ceases to have the character of a community if the interdependence of the people who live there is so slight, if their relative independence is so great that they are no longer involved in the local gossip flow and remain indifferent to any gossip control or, for that matter, to any other form of communal control. (p. xxviii)

"COMMUNITY"

Issues of informal social control are indeed central to the extensive anthropological and sociological literature on community and on small rural societies. The present state of "community studies" or the "sociology of community" has been reviewed extensively by Crow and Allan (1994) and

Harper (1989), as well as with particular reference to rural communities by Cloke (1985), Harper (1989), Saunders, Newby, Bell, and Rose (1978), and Wright (1992). With respect to St. Helena, a range of comparative ethnography can be found in studies collected in two volumes edited by A. P. Cohen: *Belonging* (1982a) and *Symbolising Boundaries* (1986a), which alongside his introductory text *The Symbolic Construction of Community* (1985), also cover a diversity of relevant theoretical approaches.

For Cohen (1985, p. 15), community denotes "that entity to which one belongs, greater than kinship but more immediately than the abstraction we call 'society' " (p. 15). However, community is not to be mistaken for uniformity. Rather, community is constituted by:

> A commonality of *forms* (ways of behaving) whose content (meanings) may vary considerably among its members. The triumph of community is to so contain this variety that its inherent discordance does not subvert the apparent coherence which is expressed by its boundaries. (p. 20)

Indeed, A. P. Cohen (1986b) considers community "as a masking symbol to which its various adherents impute their own meanings" (p. 13). Distinguishing between the "public" and "private" faces of communities, he argues that:

> The boundary as the community's public face is symbolically simple; but, as the object of internal discourse, it is symbolically complex. . . . In the "public" face, internal variety disappears or coalesces into a simple symbolic statement. In its "private" mode, differentiation and variety proliferate, and generate a complex symbolic statement.

In other words, boundaries symbolize a community to its members in two different modes, a "typical mode" and an "idiosyncratic mode." In the public face or typical mode of a community internal variety disappears, whereas in its private face or idiosyncratic mode there is a proliferation of differentiation, variety, and complexity (A. P. Cohen, 1985). Cohen (1982b) hence warns that social scientists have to be able to distinguish between "the locality's voice to the outside world, and its much more complicated messages to its own members," and to avoid "the gross simplifications of the politician, the bureaucrat, the journalist" (p. 8). Unfortunately, social scientists are frequently "fooled as much as have lay commentators into treating this simplification as an accurate statement of the community, and into using it as a formulaic key to unlock the mysteries of its culture and organisation."

However, communities may well present to their own members those very same public faces that belie the private faces of which members themselves are aware. According to Ward (1959), writing about his native St.

Helena, "there is no place in the world where everyone greets everyone else as is done by custom on this Island" (p. 430). Phillips (1986), however, does not consider such a practice to be indicative of the existence of a community in which everyone does indeed know everyone else. On the basis of his study of the English village of Muker, where people usually exchange greetings and address each other with first names or nicknames, Phillips argues that

> In this everyday practice one sees how the representational notion of the Dales community as a familiar social world wherein everyone knows everyone also serves as a normative notion: it shapes and informs people's customary behavior; it implies that everyone should know everyone and therefore should act as if they do. (p. 150)

Phillips (1986) also argues that "the idea that 'everyone's related to everyone' is a collective representation of the local community which Muker people represent to people from the outside world" (p. 143). In a seminal study of a Welsh community, Frankenberg (1957) likewise noted that although people freely admitted to having "hundreds" of relatives in the village, they were reluctant to specify their identity beyond the first cousin range.

> The belief that everyone is related to everyone else helps to maintain the exclusiveness of [the village] as a community and its unity against the outside world. At the same time, were villagers to analyze the precise nature of their interrelationships in all situations, they would reveal their disunity to the outside world. (p. 49)

The reflexive use of a community's public face is also evident in another aspect of life in small communities, namely in what A. P. Cohen (1982b), calls "equalitarianism," or "the intentional masking or muting of social differentiation, rather than belief in equality as a moral principle (egalitarianism)." That is, "the pragmatism of egalitarianism leads people to mute expressions of difference to which they are, nevertheless, sensitive" (A. P. Cohen, 1985, p. 33). In the course of his research in Whalsay, Shetlands, A. P. Cohen (1982b) found that "assertiveness of any sort, unless conducted within highly conventionalized and limiting idiomatic forms, is proscribed. . . . Even in the closest social association, relations of inferiority and superiority are rarely expressed, although they may well be tacitly recognised" (p. 11). Hence, according to McFarlane (1986), there exists a "shared idea that being 'decent' with other people requires playing down any objective superiority . . ." (pp. 97–98). In turn, "there is a general tendency to deny what are presumed to be non-merited claims to superiority by the subtle put-down and gossip. . . ." Nevertheless, given the public and private faces of communities, "such ideas and practices co-exist with quite clear ideas that hierarchies of all kinds do exist in local society and elsewhere. At most what is demanded is the playing down of hierarchical differences, a kind of polite fiction" (p. 98). As Anderson (1991) has like-wise argued with reference to "the nation," the latter "is imagined as a *community*, because, regardless of the actual inequality and exploitation that may prevail in each, the nation is always conceived as a deep, horizon-tal comradeship" (p. 98).

Given that equalitarianism is indeed no more than a "polite fiction," it is an expedient one to maintain, for "consideration of the relationships be-tween groups *within* the locale illustrates that statements of collective identity and reputation frequently provide a vehicle for the pursuit of ad-vantage amongst different community interests" (Young, 1986, p. 138). Hence, "asserting a generalised collective identity, local groups legitimise their particularistic aims by depicting them as perpetuating solidary val-ues. . . ." In this respect, equalitarianism is no more than a convenient ploy.

BACKGROUND AND SURVEY DESIGN

In order to verify the St. Helena Research Project's initial assessment of St. Helenian society, not least in the context of the above debates on small communities and informal social controls, the following sections of this chapter provide an insight into local perceptions of social and community relations through the eyes of some of St. Helena's young people. Although such perceptions may not provide a detailed ethnographic and analytical description of St. Helena, they do give expression to fundamental, yet oth-erwise implicit, notions about the character of island life; as Cullingford (1992) has shown at length, "the society they grow up in is closely ob-served by children" (p. 3). On the basis of the material presented, I sug-gest that informal social controls, such as gossip and equalitarianism, are indeed integral to the character of social relations on St. Helena, as are the contrasts between the public and private faces of the island's community.

Research into the perceptions, attitudes, and aspirations of children and young people on St. Helena has been carried out by Charlton and O'Bey (1997), R. Cohen (1983), Gillett (1979), a team from the Overseas Development Administration (1993), and Royle (1992); yet none of this research explored the complexities of local experiences and perceptions of community, not least with reference to informal modes of social con-trols. Instead, the character of St. Helenian society and culture has largely been taken for granted, not least because of long established Arcadian imaginings of St. Helena as the "Island of the Blessed" (Schulenburg, in press).

On the basis of previous St. Helena field research, which I carried out in 1993–1994, it was decided that perceptions of community could be best elicited by asking pupils at St. Helena's only comprehensive school, Prince Andrew School, to write essays on the topics "The Community in Which I Live" (Year 8) and "People on St. Helena" (Years 9 and 10). Both R. Cohen's (1983) and Gillett's (1979) research employed essays to good effect. Given that pupils were to be made aware that these anonymous essays would be evaluated by a nonislander, it was assumed that they would feel free to express their views of their fellow islanders—views that would not necessarily be verbalized, let alone be put in writing. This did indeed turn out to be the case, although it was found that the essays on community primarily provided descriptions of the geographic rather than the social dimensions of a given pupil's habitat. In consequence, the statements drawn on in this chapter are taken solely from the essays entitled "People on St. Helena."

The essays that formed part of this study were written during normal class hours in December 1998. The exercise was set by teachers in one class in Year 9 and one class in Year 10. No guidance was given on the completion of the exercise. At the time of the exercise, the total number of pupils in the years taking part amounted to 81 pupils in Year 9 (ages 14–15) and 78 pupils in Year 10 (ages 15–16). A total of 41 essays were returned, representing the number of students in these two classes on the days in question, which in turn constituted 22% of the nominal number of pupils in all classes in Years 9 and 10. Although they had not been requested by me, an additional 6 essays were returned by pupils in one class in Year 11 (ages 16–17); this was 7% of the nominal number of 84 pupils in that year (the remainder of their classmates had been occupied with another research exercise). As there was no marked difference in style or content, all these essays were grouped together. In any case, no information on the age and gender of respondents had been requested. There is no reason to think that the essays are unrepresentative of pupils in the age groups in question.

Rather than reproducing the essays in the form in which they were written, they have been divided into separate statements that were subsequently grouped under a number of generic headings. Upon reading the essays, certain themes began to emerge that formed the basis of the categories to which particular extracts have been assigned. Within each category an attempt has likewise been made to group related statements with one another and to impose some sort of progressive order on the whole. Although the essays are not reproduced in their entirety, only those statements have been omitted that were virtually identical not only in sentiment but also in the terms in which these sentiments were expressed. This was done solely to keep this chapter within manageable proportions despite the loss of a limited quantitative dimension. Also, for the present purposes, only those statements have been reproduced that had a bearing on the topic of this chapter, particularly on the character of small communities and informal social control, although a more exhaustive treatment of these essays can be found elsewhere (Schulenburg, n.d.). It is worth noting that there was no need to alter any of these essays in order to preserve anyone's anonymity, although given the circumstances of this study it was impossible to preserve the anonymity of the school from which the pupils were drawn or the community of which they are a part. Last, each essay has been assigned a number (shown in square brackets) to allow the reader to identify and relate to each other the remarks of a given respondent, while the idiosyncratic spelling and grammar of the originals have been maintained.

On what basis, though, can it be assumed that pupils provided honest accounts of their views? First, given my familiarity with St. Helena on account of extensive field research undertaken prior to and in connection with this study (13 months in 1993–1994, and 4 months in 1998–1999), I believe that the portrait of St. Helena which emerges from these essays is consistent with my own and other observers' (e.g., UNDP, 1999) assessment of the character of St. Helena society. Any deliberate exaggeration or falsification of fact would have stood out plainly. Second, rather than writing dishonestly, pupils who chose not to cooperate with this research simply chose to write little more than a couple of lines. All told, pupils had nothing to gain from misrepresenting their views.

SOCIAL AND COMMUNITY RELATIONS ON ST. HELENA

Drawing on a total of 47 essays on "People on St. Helena," this section provides extracts grouped under the following headings: friendliness, community, cooperation, social services, nosiness and gossip, conflict and division, crime, and young people.

Friendliness

The most notable defining characteristic of St. Helenians was seen to be their friendliness and good manners, a point made by a large number of respondents.

[9] People on St Helena unlike other countrys or place if you walk pass anybody they normaly will say hello.

[36] People on St. Helena are friendly, caring and loving. They have respect
for you when they pass you on the road.

[40] The people of St. Helena has a lot of maners and will only speck to you
when you speak first, but are very friendly.

[34] The people on St Helena are friendly, kind and to others.

This friendliness is acknowledged despite reservations:

[38] The people on St Helena are friendly but some can be snobbish. [. . .]
In the shops they are very helpful and always smiling. Also the people
are caring and kind.

The friendliness of islanders is identified in particular with their attitude
shown to tourists and visitors, that is, outsiders to island communities.

[21] The people on St Helena are very friendly and that is what most tourists like about the island. When St Helenians see tourists, they welcome them with open arms. Even if they don't say 'hello', the Saints
smile or wave to whoever they see. This tends to bring about a calm
feeling on the island.

Pupils know that tourists and visitors are well aware of this friendliness
shown towards them and that they value it.

[30] most people who come here from overseas feel very welcome by the
people of St Helena, and is told they felt really warm welcome by
saints.

Nevertheless, these remarks are once again qualified.

[36] Other people that come to visit the Island, often say we are friendly,
but this does not go for all the Saint Helenian's.

 [6] On St Helena, the people can be very bad mannered, in speech and
ways.

[13] Some people walk pass each other without speaking.

Community

Despite such reservations, the friendliness of St. Helenians is seen to be
reflected in the close-knit nature of island community.

[26] My Opinion of the People on St Helena is that they are unique and different to others of this World. By saying this I mean that they are as if a
family, all cooperating as one, getting on with one another.

[33] People on St Helena have a lot of close releiships [*relationships*]
where ever you go you see them having private times.

[41] almost everyone has a relative on the Island

[23] Everyone like a big family at our high school.

[22] At school children are bounded together. They form groups and everyone does get along with everyone.

There is one obvious reason why this should be so.

[21] St Helena is a very small island and that is basically why everyone
seems to know everybody.

[22] Due to the island being very small, everyone basically knows almost
everyone even if they do not associate with them.

There are indeed reasons why St. Helenians may not, after all, associate
with one another despite living in such a close-knit community.

[23] St Helena is a very small island. Everybody knows everyone. Even if
they don't really know them they still speak to the people they don't
know. If your walking down the street and you see someone you dont
know you will out of habit speak to them. On the other hand, if you really don't like a person you make it your business not to speak to them.

[36] Although St Helena is a small Island, not every one on St Helena know
each other. A great many of them don't even like each other.

Cooperation

Friendliness and a sense of community appear to go hand in hand with a
local spirit of cooperation and mutual assistance.

[21] Because most of the houses are quite close together on St Helena, everyone treats everyone like their neighbour.

[30] Saints have their own opinion of different things but are willing to help
one another in any way that they can.

[8] People here also likes to help others and they are very understanding
and consider others, in need of help.

[32] They all ways appreciate whatever someone else does, because some
families doesn't have water or electricity.

[33] People on St Helena they share lots of there fruit and there vegetable
to other family and friends. They have people on St Helena which have
great care for the very old and sick people even if the people hardly
know them by there name or character they are willing to help them all
who is done up and old/sick. The children, young children have great
care for there granparents/old people in there family.

[22] At this present day there are no home-less people on St Helena or children without a family which goes to show that everyone does care for
each other.

[35] In other places of the world its is hard to trust people, but on St Helena you can trust a lot of people.

Social Services

Local government too is seen to have a share in fostering this spirit of cooperation, primarily in the provision of social services.

[23] We have one hospital and one place for the mentally disabled, one
place for the handicaped and one place for the older people who can't
take care of themselves.

[21] If children are unwanted then there is a Children's Home for them.

[35] the people on St Helena who doesen have any job are provided with
money from the government.

[36] the people on St. Helena depend on Social Service for money if they
are not working.

[14] the social services will provide food and clothes for children and
grownups so people are not always greedy.

Nosiness and Gossip

Alas, a close-knit community also entails attitudes and behaviors that run
counter to the friendly and caring attributes of St. Helenian society.

[39] Although people are friendly they sometimes can be nosy and mind
other peoples business.

[14] People on St Helena are sometimes nosey (grownups) and always like
to gossip about other people.

[23] If you don't go out on the weekend you can get on the phone and find
the gossip.

[1] On this island the way it seems is that whatevers your business is everybody elses.

[10] Others are very nosey and don't mind their own business and call people names, but don't look at anyone else how bad they are in their own
family.

Such behavior, however, is seen to be more than a mere pastime but is
considered to have serious implications for young people's lives.

[6] Many of the people can be very nosey, and mainly the grown-ups like
to talk about other people's children instead of worrying about their
own.

[10] People on the island sometimes over-react especially our parents, whenever they learn that someone saw you with a boy or just having fun (which they call bad behaviour) they go and tell your parents, mainly something that isn't true and they get you in trouble and be given a bad name.

[1] Another thing about this island and its people is that when someone sees a girl with a boy just having a bit of fun they will go tell your parents that you were doing something that you didn't and this gets you in trouble.

This is a problem affecting adults as much as children.

[8] Some people also find it hard to find a job because of the bad reputation they has.

Interestingly, the causes of such nosiness are believed to be the same as those that account for the perceived sense of community on St. Helena.

[6] Been [? *Seeing*] as the island is very small, and there's nothing else to talk about, so the grown-ups (meaning, mainly women) like to interfare with [*i.e., comment on*] other womens clothing, but only because they are jealous, of not having as nicer clothes as the person they are interfaring with.

Conflict and Division

Not surprisingly, children note that aside from nosiness and gossip, there are factors that account for St. Helenian society being characterized other than by a spirit of cooperation.

[8] People here donot keep together as one big family they all keep with their familys and friends.

[1] There is a lot of competition on this island, like everybody wants to be better than each other and I think this is very childish.

[10] There is a lot of competition on St. Helena, e.g the clubs in entertainment, everyone like to beat each other, also the builders each one likes to have more workers and earn more money than the next.

A particular problem is seen to be what St. Helenians call "big ways," or an inflated opinion of oneself.

[18] My opinion of the St. Helenian people are that some have no manners and got big ways. The trouble is that some of them smoke and drink so much they don't even know whats going on.

[19] The peouple on St Helena they are that some people get big ways like the police

[20] Some what work for SHG [*St. Helena Government*] got big big ways

[15] There are people on St Helena who want's you to respect them but they will not respect people in public.

Lastly, conflicts on St. Helena can even get physical.

[31] Some st [*? Saints*] are bad because the is lots of frights [*fights*] on the island.

[27] Some people on of St Helena who like looking for trouble and some don't and the people like looking for trouble always find of the results like when they fights and come out with brues on them and scars.

Some children consider such behavior, including big ways, to be typical of certain of the island's districts.

[17] The Longwood people is the crazies people going, because they like to treated people like saying are going to beat them up that's why they called that place 'TEXAS'.

[20] some people from James Town has big ways and big head[;] some people from Longwood is cool

Crime

Pupils' opinions of crime on St. Helena likewise show that some people's attitudes leave something to be desired although, as expected, opinions do differ.

[23] We have one small police force we don't really need a big one because to my knowledge there isnt much trouble.

[33] There arent dangerous people on the Island like in other places.

[38] There isn't always good people on St. Helena, there can be people who do naughty things.

[13] Some people on St Helena are very naughty. Some people steel and so are very shy.

[31] same staints [i.e., *Saints*] try to bring drugs to the island.

[8] A very few of the St Heleians go to jail for a minor offence nobody have ever committed a crime.

[23] On St Helena there isnt much law breaking, you won't find children taking drugs. You won't find children in prison.

[21] If there is underage sex, drinking or smoking on the island, then the people concerned have to go to court.

[35] People on St Helena are less violent and dangerous, but some time
      people cane be dangerous and break the law.

Despite any such incidents, one pupil asserts that:

[21] Most people on St Helena are peaceful and there is rarely much vio-
      lence.

The explanation for the acknowledged low rate of crime is found in aspects of island society that have already been mentioned earlier.

[22] It is probably because of this friendship that crime rates are not as high
      as they may be else-where in the world.

Young People

Pupils also commented extensively on the concerns and lives of young
people themselves, not least on some of the problems and divisions they
encounter.

[1]   Some of the people on this island mainly some of the children criticise
      others, they call them names and sometimes there is no reason for this.
[11] so of the teenage boys are womeny. They keep with girls most of the
      time anyway. Some of them like making enemys and making fun of
      each other and calling people names.
[22] The childrens' overall behaviour is in my opinion better than that of
      other children in the world. They do associate with each other at
      school or saturday nights but drug abuse is not to my knowledge a part
      of their lives.

Misdemeanors and their consequences are a topic of particular concern.

[23] There is alot of drinking alcohol, smoking cigarettes and sex under age
      (16) But everyone concerned is so smart about it nobody gets caught.
      Abortions do take place But not very many.
[21] If a woman become pregnant then she can have an abortion if she
      wants to.
[31] Some people take part in under age sex because they are attack [at-
      tracted] to the oppersiat sex.
[6]   Now as the girls of the island is growing up, some of the teenagers, are
      starting fights, and doing things that only grown ups should be doing.

Pupils acknowledge that their behavior is not appreciated by all, although they differ on whether and how to take account of this.

[1] The people on this island sometimes overreact mainly the parents, they get all angry and upset over nothing and everything is 'your still under age'. I know that they are just looking out for you but they look out for you too much, and when you do something wrong the word 'GROUNDED' is then mentioned.

[13] Some people on this island do not respect other people who were older.

[5] The teenages on this island is getting out of hand but actually I don't blame them.

QUANTITATIVE CONTEXTS

According to Cullingford (1992), qualitative data of the kind presented above do not merely constitute "a series of anecdotes but an overall picture of a frame of mind shared by the children" (p. 11). Hence, although such an overall picture "does not lend itself to statistical sophistication, . . . it does underlie the essential truths" of children's lives (p. 13). It is also the case, as Lambert (1968) argued, that children's views of their experiences do not constitute an "objective evaluation" of those experiences. A child's or a young person's perspective, with its "own slant and narrowness," may give little expression to some aspects of life, yet "incomplete as is their perspective, and raw as is their judgement, it provides a penetration, a sensibility and a freshness of its own" (p. 6).

Statements such as those by St. Helenian pupils quoted in this chapter are of particular value, according to Brewer et al. (1998, p. 575), in that they allow for a focus on "people's accounts of their experiences of local community," thus obviating the need to "operationalize community structure in terms of objective statistical measures," which "inadequately reflect variations in social bonding in small localities" (also see Hawtin, Hughes, & Percey-Smith, 1994). Yet, it would surely be of interest if one were to compare these young people's views of their society with appropriate quantitative data from St. Helena. As the *St. Helena Diocesan Magazine* asserted as long ago as 1925:

> Of our social life, superficially, there may be quarrels and disagreements as in every other community all over the world, but look deeper and you will find that we are a wonderfully contented and peaceable people. Actions for divorce, libel, scandal, assaults, are practically unknown. (p. 68)

Although the scope of this present chapter does not allow for a thorough comparison between the statements made by St. Helenian pupils and available St. Helena statistics, a brief examination of such data in the area of crime does however serve as a useful exemplar. This is instructive not only because of current concerns with the relationship between television, behavior and violence, but also because there is a common assumption in the words of one pupil, that "[21] most people on St. Helena are peaceful and there is rarely much violence."

According to police statistics (St. Helena Government, SHG; 1997b), a total of 212 crimes were "reported" on St. Helena in 1997 (i.e., 0.042 per inhabitant on the basis of a population of 5,010; *St. Helena Census*, 1998). Of these, 32 offenses were reported under the Offences Against the Person Act 1861; these do not include assaults against police officers (i.e., 0.005 per inhabitant), although a total of 96 offenses were reported under the Criminal Damage Order 1979 and the Theft Act 1968 (i.e., 0.019 per inhabitant).

Interestingly, although offenses against the person make up only 15.09% of all reported crime on St. Helena, according to a recent St. Helena Government survey (1997a), 38% of persons taking part in the survey and who have had reason to contact the police, did so concerning "a crime involving violence" (p. 13). Although the above figures would appear to confirm that crime on the island is indeed lower than elsewhere, reported and recorded levels of crime on St. Helena are known to be well below their actual levels. As that same survey found in response to a series of questions respecting people's reactions to witnessing crime, "a high proportion of respondents would do nothing in most situations. This is probably due to the fact that there are small communities and no one wants to get involved as being a witness etc." (p. 8). A recent United Nations Development Program (UNDP) report on St. Helena (1999) similarly found that "there is unwillingness by the public to get involved in criminal investigations due to the fact that everyone knows everyone else and people have to continue to live with each other" (p. 43). Hence, in small communities informal social controls not only have the effect of deterring crime but also of deterring the reporting of crime and the prosecution of offenders.

With respect to the debate on television, behavior, and violence, the above statistics of levels of crime on St. Helena are also of interest insofar as according to Gunter and McAleer (1997, p. 100), one of the most powerful factors influencing viewers' perceptions of television violence is what they call "familiarity of surrounding," that is, the "closer to everyday life the violence is portrayed as being, in terms of time and place, the more serious it is judged to be." Gunter and McAleer (1997) furthermore report that research has shown that "television portrayals of violence not only taught aggressive behavior, but also demonstrated who is most likely to fall victim to violence, thus engendering enhanced fear of violence among certain categories of viewers" (p. 93). Intriguingly, given the perception of St. Helena as crime free, a recent survey about public attitudes to policing and fear of crime (SHG, 1997a, p. 7) determined that 30% of women on St.

Helena were "very worried" about being raped or sexually attacked, whereas 6% were "fairly worried." Only 46% were "not very" or "not at all worried." Likewise, 48% of men and women were "very" or "fairly worried" about being assaulted. This compares to the fact that 91% of men and women respondents said they had never actually been the victim of "violence or abuse" (p. 5).

Based on an analysis of quantitative data along the lines of the above exemplar, it may well be found that islanders' accounts of the nature of St. Helena society do indeed differ from the picture that results from quantitative indices, especially if the former overemphasize a community's positive aspects. However, positive accounts must not be considered as false or even deluded for any such representations have an important normative role to play (Little & Austin, 1996; see also Holy & Stuchlik, 1983). Insofar as pupils on St. Helena view their fellow islanders in a predominantly positive way and value their society for its positive qualities, these views inform islanders' conduct by setting common standards for social interaction. With respect to these norms, it could indeed be argued that St. Helena constitutes a generally "healthy community environment." The acceptance of such a characterization, however, should not mislead one into ignoring the less "healthy" aspects of island society. Neither should the positive nature of community be taken for granted, as if the term did not need qualification (Little & Austin, 1996; see also Darcy, 1999).

CONCLUDING REMARKS

The arrival of broadcast television on St. Helena and its incorporation into daily life may well have a welcome effect on the intensity of surveillance in St. Helena society, above all by diverting the attention of "watchers" toward watching television and away from one another (Morley 1992, p. 152), as well as by creating new "back regions" (Giddens, 1985; Goffman, 1969) or by enhancing existing ones. Although, according to Morley (1992, pp. 224, 237), "home-based consumption represents a retreat from the public realm of community," as "the private individual retreats into his (or her) house and garden," the "domestication" of communication technologies has also "increased the attractiveness of the home as a site of leisure. . . ." Back regions on St. Helena are consequently enhanced by the fact that the domestic sphere "has come to be centrally defined as the social space within which individuality can be expressed—the refuge from the material constraints and pressure of the outside world" (p. 225). Television and associated home-based information and communication technologies may well provide welcome relief from the pressures of St. Helena's "watchful" society, even if at the cost of the more positive aspects of life in a small community. Tomlinson (1997) notes that "fear is already being expressed that television will begin to erode the special characteristics of St. Helenian culture and society" (p. 189).

Although radio and television are "mechanical extensions of the experience of going places and meeting people," enabling listeners and viewers to "visit" more places and "meet" more people, "these experiences need not replace the village club and the Church choir nor the friendships made in them" (Bracey, 1959, p. 201). Hence, according to Falk (1998), it is "naive to assume that the advent of the new communication options of the web will necessarily lead to less face-to-face communication." Indeed existing communities, especially diasporic ones such as St. Helena, "may be reinforced through the use of new communication technologies" (pp. 290–291).

The widening and opening up of previously more confined communities on account of new communication and information technologies (e.g., "cyber communities"), may well lead to the additional creation of so-called weak ties. As Jacob, Bourke, and Luloff (1997) have found, one of the drawbacks of small or rural communities is that they feature fewer weaker ties on account of the lack of extensive social networks and a limited number of possible contacts. "Weak ties are secondary and tertiary acquaintances that are not especially intense, but comprise an important part of an individual's support system" (p. 277). Evidence for such support can be found in studies by Luloff and Swanson (1990) and by Zekeri, Wilkinson, and Humphrey (1994). Given the relative absence of weak ties in a community as small and isolated as St. Helena, "the implications of islanders becoming hooked into the global media network" (Tomlinson, 1997, p. 189) may well be beneficial ones.

There is, however, evidence to suggest that factors other than new information technologies have already had a key role in changing the notions and attitudes of St. Helenians. This is attested by a once confidential government report (St. Helena Government, 1980), which concluded well before the arrival of television that "wants and expectations have recently been growing amongst islanders far in excess of the wealth of their island," not least on account of "employment on Ascension which influences even those who do not go there by the inflow of goods bought with money earned there" (p. 1).

According to a recent paper by Basil George (1998), an islander and former chief education officer, there are troubled times ahead on St. Helena but not because of any supposed ill effects of television.

> The current economic situation of St. Helena is such that many people are leaving and many more would leave if there were the opportunity. . . . In-

creasing the number of unemployed increases the social problems which is borne out for example by the rise in burglaries over the past few years. . . . The island also has a rising number of elderly people. Many of the population who would normally take care of relatives are working offshore. Though income from this source is the mainstay of the economy, it disrupts family life and is eroding the fabric of island society. . . . The present social and economic climate means that with so many people absent from the island, traditional family support is beginning to disintegrate. . . . The number and age groups of people remaining in the human habitat of St. Helena to sustain services and accept family and social responsibilities are reaching a crisis point.

These views are confirmed by the recent *Human Development Report for St. Helena* (UNDP, 1999), which was coauthored by both islanders and expatriate officers. Despite its many flaws, this public report is noteworthy not least for presenting an usually candid picture of "community, the family and social issues" (the title of one of its chapters), thereby countering prevailing representations of community life on St. Helena and addressing a number of rather sensitive issues including domestic violence, alcohol abuse, sexism, and racism. Significantly, especially within the context of this chapter, the report concludes that:

> Perhaps the most striking feature of family life in St. Helena, is the fact that so many households are missing significant individuals because they have left to take jobs overseas. . . . There is certainly some evidence in the schools, that children's behavior is adversely affected by the disruption in their family lives. (p. 37)

In his interim report on the impact of television on St. Helena, Charlton (1998a) was hence right to note St. Helena's "climate of considerable social, political as well as economic change" and to conclude that "the processes linked to—as well as the outcomes of—these changes are liable to make unprecedented impact on family and community life on the island" (p. 32). In 1994 the possible knock-on effect of this state of affairs had already been pointed out by George (1994) in an article on the then low rate of recorded juvenile crime on St. Helena.

> Good behaviour is a quality that we must recognise and appreciate in our young people. It is reassuring to know that this traditional value is still with our children as they grow up in St. Helena. We need however, to work at keeping these qualities in the family. It is not going to be easy. We cannot remain isolated from the influence of the outside world. . . . Education makes an important contribution in carrying forward traditional island values and

in preparing people to accept and adapt to change, though it acknowledges
that parents and the family play the key role. (p. 6)

Despite the pessimism expressed by some, it would appear that such
"traditional values" will indeed survive, as the arguably beneficial, if also
restrictive, aspects of social and community relations on St. Helena are not
peculiar only to St. Helena. Instead, they are an integral aspect of life in
small communities in general, and as such, are unlikely to be subject to
any significant degree of change.

ACKNOWLEDGMENTS

I gratefully acknowledge the cooperation of the pupils and teachers at
Prince Andrew School in St. Helena, and the advice on the literature that I
received from Roy Dilley, Grace Barnes, Rene Bekkers, Kevin Meethan,
Thomas J. Scheff, and Richard Wilsnack. Tony Charlton and Andy Hannan
provided extremely helpful comments on earlier drafts of this chapter.

CHAPTER THREE

The Nature of the Television Output

Barrie Gunter
University of Sheffield

The principal objective of the St. Helena project is to investigate the impact of broadcast television on the island's children. As the previous chapter explained, television's impact in this context has been measured in terms of observable and reported changes in children's use of time, leisure pursuits, and social behavior. Two broad perspectives to measuring television's effects have been adopted in this research. The first has focused on the impact of the presence of the new medium per se by comparing behavioral changes among children before and after the introduction of satellite broadcast transmissions to the island. The second has been concerned with the effects specific types of content within programs might have upon selected behaviors.

The effects of television on out-of-school activities, for example, have been assessed primarily in terms of readjustments the children might have made to the way they use their time following the introduction of broadcast television to their community. Here, it is assumed that because watching television uses up time that time must be subtracted from the child's daily time budget. Unless children normally have "spare" time that is unoccupied, one can expect a reduction in the allocation of time to other activities to accommodate the need to watch television. Behavioral changes of this sort are hypothesized to be content independent. This means that they are linked to the introduction of television, rather than to the consumption of specific types of program. With regard to the effects television might have on children's social behavior, however, attention shifts to the nature of television's broadcast output. The interest in this context centers on changes to patterns of overt behavior that could hypothetically be linked to a child's identification with behavioral examples depicted on screen. Behavioral changes here are therefore more likely to be content dependent. To explain these types of effect, one must understand the way children use television and the kinds of programs they typically choose to watch.

In order to assess content dependent effects of broadcast television, it is vital as a first step to conduct an analysis of the nature of that content. Television programs represent many different genres. Different genres have different functions for audiences and, consequently, contain different types of subject matter. As with broadcast television in many other parts of the world, the television services transmitted to St. Helena are designed to provide information and entertainment. Some of the services are designated as being dedicated to the entertainment of child audiences and others are designed primarily for adult audiences. It was regarded as an essential part of the research project that an analysis should be conducted of television output. The main reason for this was so that an examination of content-dependent effects of television on children's social behavior could be informed by a detailed account of the nature of children's viewing diets. The main focus of attention was placed on the measurement of antisocial behavioral effects of television, which meant that an analysis of program content relevant to such effects would be centrally important. In particular, any such analysis should aim to establish the quantity and nature of depictions of violence in programs broadcast to St. Helena. Within 3 years of the onset of broadcast television, therefore, a content analysis study was instigated to assess the portrayal of violence on the island's television services. This chapter presents the key findings of this study.

THE SIGNIFICANCE OF TELEVISION VIOLENCE

For nearly half a century, the analysis of violence on television has been a major area of research enquiry. The establishment of how much violence is shown on television has been regarded as an important matter for two main reasons. First, there is a common belief regularly promulgated by critics of the medium that the general public regards the depiction of violence on television as excessive and distasteful (Gunter & Stipp, 1992; Gunter & Wober, 1988). Second, a large amount of published scientific evidence has indicated that televised violence can cause adverse short-term audience reactions and the development of longer-term patterns of antisocial behavior. Such effects were deemed to be especially likely to occur among children who are regarded as more vulnerable than adults to television influences (Bandura, 1994; Friedrich-Cofer, 1986; Huesmann &

Eron, 1986). Although this evidence is not entirely unequivocal (see Lynn, Hampson, & Agahi, 1989; Wiegman, Kuttschreuter, & Baarda, 1992), the literature on audience research does provide relevant input to any discussion about the classification of violence on television—a topic that is central to the work discussed in this chapter.

Even during television's early days, portrayals of violence were identified as regular aspects of prime-time programming (Remmers, 1954; Schramm, Lyle, & Parker, 1961; Smythe, 1954). The quantification of violence on television has taken place around the world revealing that violence is a popular feature of entertainment programming across many different cultures (Bouwman & Stappers, 1984 in The Netherlands; Caron & Couture, 1977; Cumberbatch, Lee, Hardy, & Jones, 1987; Ginpil, 1976; Gordon & Singer, 1977; Gunter & Harrison, 1997a,b; 1998 in the United Kingdom; Haines, 1983; Halloran & Croll, 1972; Iwao, de Sola Pool, & Hagiwara, 1981 in Japan; Linton & Jowitt, 1977; McCann & Sheehan, 1985 in Australia; Newell & Shaw, 1972; Shinar, Parnes, & Caspi, 1972 in Israel; Stewart, 1983; Watson, Bassett, Lambourne, & Shuker, 1991 in New Zealand; Williams, Zabrack, & Joy, 1982 in Canada).

The established methodology for this work has generally adopted a standard quantitative analytical approach that utilizes an a priori definition of violence enabling trained coders to identify on-screen occurrences of qualifying behavior (see Halloran & Croll, 1972; Gerbner, 1972; Gerbner & Gross, 1976; Gerbner et al., 1977, 1978, 1979, 1980; Gunter & Harrison, 1998). In addition, a coding frame is supplied that provides further specifications concerning units of measurement and the characteristics of the depicted violence that are to be cataloged. The ways in which this methodology are applied can vary. Some researchers have studied all types of television output, whereas others have been more selective, focusing on specific genres or day parts to the exclusion of others.

In many countries with long-established television services, violence has typically been found to occur in more than half the programs broadcast on mainstream channels and particularly during those hours when the largest audiences are watching (Cumberbatch et al., 1986; Gerbner & Gross, 1976; Gunter & Harrison, 1998), whereas in some countries the proportions of programs with violence are considerably greater than that (Haines, 1983; Iwao et al., 1981; Menon, 1993). In recent years, the emphasis has shifted away from simply quantifying violence on television to a closer examination of the nature of that violence and the types of programs in which it predominantly occurs (Gunter & Harrison, 1998; Potter et al., 1996). This revised perspective recognizes that violence can take on many different forms and that audiences respond in different ways to violent scenes. The nature of viewers' responses can depend critically on the form of the violent behavior, the nature of its consequences for victims, the type of aggressor and victim, and the context or setting in which the violence occurs (Greenberg & Gordon, 1972a, 1972b, 1972c; Gunter, 1985; Kunkel et al., 1996; van der Voort, 1986). Some writers have discussed "risk factors" in relation to on-screen portrayals of violence. These are attributes such as realism of setting, reward and punishment of actions, motives of perpetrators, characteristics of victims, and form of aggression that have been previously shown to mediate viewers' subsequent perceptual, emotional or behavioral responses to screen depictions (Potter et al., 1996; Wilson et al., 1996). A more informed account of violent content is believed to derive from coding frames that incorporate such elements in their analysis.

CENTRALITY OF CONTENT ASSESSMENT
TO TV EFFECTS

Most of the research into the occurrence of violence on television has taken place in communities in which television was an established feature of the social environment. Only a handful of research studies have taken place in communities for whom television was a new phenomenon (Brown, Cramond, & Wilde, 1974; Furu, 1962; Granzberg, 1982; Granzberg & Steinbring, 1980; Himmelweit, Oppenheim, & Vince, 1958; Schramm et al., 1961; Williams, 1986). In most of these cases, research was conducted on the displacement by television viewing of other activities and the impact of television on intellectual development and social behavior. The key weakness of much of this earlier work stemmed from the tendency to consider television as a whole as the main intervention, rather than to investigate the differential effects that might follow from exposure to particular types of program content. However, in order to understand more fully the effects that television might have upon its audiences, especially children, one vital ingredient concerns the nature of the program output being beamed to the community under investigation.

Modern-day television services can broadcast a wide variety of content to audiences, the effects of which cannot all be placed within a single category or assumed to operate in a single direction. In the context of the impact of televised violence, for instance, adopting the mere presence of television as the key treatment variable in a field study does not indicate how much the medium is used by viewers in a community, or the different ways in which individuals use the medium. When measuring content-dependent effects, it is essential to have detailed data on individual viewers' television exposure patterns and about the content of their viewing diets. Thus, the children under investigation on St. Helena kept television viewing diaries to indicate over specified time periods which programs they watched. In addition, content analysis was conducted over those same time periods to establish how much violence, and of what kinds, these programs contained. It was then possible to weight the children's viewing by the content analysis data and obtain precise measures of their level of exposure to televised violence. The results of the latter analysis are reported in chapter 5. The remainder of this chapter will discuss the violence profile of St. Helena television.

ANALYZING VIOLENCE ON ST. HELENA TELEVISION

Content analysis was carried out on small samples of broadcast television output over 2 years—1997 and 1998. The programs were taken from the five television services beamed into the island: CNN (news), Hallmark (films), Discovery Channel (documentaries), SuperSport, and the Cartoon Network. The research adapted a methodology used in an earlier investigation of violence on British television (Gunter & Harrison, 1997, 1998).

Program Samples

The programs were video recorded from 3 days of television output each year (Sunday, Tuesday, and Thursday of 9, 11, and 13 February 1997 and Sunday, Tuesday, and Thursday of 29 and 31 March and 2 April 1998). The total coded program samples comprised 153 programs in 1997 and 136 programs in 1998. These program samples aggregated respectively to 79.5 and 84 hrs of broadcast output that were available to be monitored. Video recordings were made in two spells (6:00–9:00 a.m. and 3:00 p.m.–12:00 a.m.) on the 2 weekdays and for one continuous spell (6:00 a.m.–12:00 a.m.) on Sundays. The weekday recordings covered the broadcast hours available to children before and after their school day.

Coded programs were classified into 27 different genres that were clustered into five superordinate genres: drama, children's programs, sport, factual, miscellaneous (entertainment and music, arts, and religion). Advertisements were not coded. Four broad categories of programs predominated: drama (including cinema films and drama programs from the UK, USA and elsewhere); children's cartoons; sports (live and highlight coverage); and factual programs (including news, current affairs, documentaries, and magazine programs). The distribution of monitored programs by genre is shown in Table 3.1.

Definition of Violence

Violence was defined as

TABLE 3.1
Distribution of Monitored Programs by Genre

	1997		1998	
	N	%	N	%
Children's cartoons	37	24.2	32	23.5
Sport	38	24.8	36	26.5
Factual	68	44.4	58	42.6
Drama	13	6.6	9	6.6
Miscellaneous	—	—	1	0.8
Total	153	100.0	136	100.0

Any overt depiction of a credible threat of physical force or the actual use of physical force, with or without a weapon, which is intended to harm or intimidate an animate being or a group of animate beings. The violence may be carried out or merely attempted, and may or may not cause injury. Violence also includes any depiction of physically harmful consequences against an animate being (or group of animate beings) that occur as a result of unseen violent means.[1]

---

[1]Further stipulations and considerations were presented to coders to guide their cataloging of violence. An *Overt Depiction* means that the violence is occurring on the screen. The threat, the act, or the harmful consequences are shown or heard in the action of the plot. Verbal recounting of previous threats or acts of physical force or talking about violence does not count as violence. *Physical force* refers to action intended to cause physical pain or injury to an animate being or the use of physical tactics (such as strong-arming) that are intended to coerce the action of another in a way that threatens harm. Physical force must be enacted by an animate being; physical force or harm that occurs as a result of an act of nature (e.g., an earthquake) is not counted as violence. Physical force can be perpetrated against either the self or another.

A *Credible Threat* means that within the context of the plot the perpetrator must display an intent to harm. Joking threats that have no believable intent to harm are not included (e.g., a character who is embarrassed and says "I could kill you for saying that"). An *Animate Being* refers to any human, animal or anthropomorphized creature (non-human) that possess attributes such as the ability to move, talk, think, act against something. A group of animate beings can involve several creatures or more abstract collections of creatures such as institutions or governments. Violence must involve at least one human or anthropomorphized being, either as perpetrator or as victim. An act of nature involving two or more animals that threaten or harm each other is not considered violence so long as the animals are not anthropomorphized (i.e., possess human-like characteristics like talking, thinking, etc).

Accidents that involve physical harm are considered violence *only* when they occur in the context of an ongoing violent event (e.g., police chase a robber who accidentally falls off a building during the pursuit) or when they involve the use of weapons (e.g., two children are playing with a gun that goes off and hits one of them). Physical force against **Property** (e.g., breaking a window; setting fire to a building) will be considered violence **only** when such actions are directed at intimidating, punishing, vandalism, or seeking revenge against an animate being. Property damage due to accidents, even in the context of an ongoing violent action, will not be considered violent.

The coders were given a series of further instructions in respect of defining and applying the basic currency of measurement—the "violent act." A violent act was defined in three ways. Coders were instructed not to attempt to count every single blow that was struck or shot that was fired. Thus, as a unit of measurement, a violent act might involve the same repeated behavior by the same perpetrator (i.e., A punches B five times equals one act of violence because A commits the same violent action several times). A change of violent act was defined by a change of perpetrator, a change in the form of violence, or a change of scene. Thus, if A hit B and then B hit A in return, such a scene would represent two acts of violence; here A commits one violent act and then B commits another violent act. Another distinguishing feature would be any change in the nature of the behavior itself. Thus, different actions by the same perpetrator might represent more than one act of violence, even if they occurred in an unbroken sequence. For example, A punches B and then immediately afterward, also kicks him. This would represent two distinct acts of violence, an act of punching, and a separate act of kicking. Finally, a change of scene would also serve as a means of dividing up violent acts. If A hit B and then the camera switched away from this scene for several seconds to a nonviolent scene before returning to A still hitting B, this would be counted as two separate violent acts even though the perpetrator, victim, and form of violence had not changed.

Coding Schedule

The coding schedule contained a series of questions about each violent act. Initial information was coded for each act of violence concerning the program in which it occurred, the genre of that program, the channel on which it was broadcast, and its transmission time. The length of the program and of the violent act in seconds was also noted. A number of specific features of the violence and its setting were then cataloged. These features included the historical setting of the action, the country where it took place, the motivational context of the violence, the injury consequences of violence, the form of the violence, and type of weapon used. Finally, coders also noted the gender, ethnic origin, age, and occupational and dramatic role type of aggressors and victims in each violent act, and for the aggressors, made a note of their individual goals or motives for being violent.

Coding and Reliability

Coding was conducted by teams that comprised five coders in 1997 and six coders in 1998, aged 21–50 years. All coders in 1997 and five in 1998 were female and all but one were educated to graduate level. Reliability checks were run to ensure that coders used the coding schedule in a consistent fashion. In both years, coders participated in a series of training sessions before the coding exercise began. These sessions were designed to familiarize coders with the details of the coding schedule and to ensure they understood the definition of violence, the instructions for distinguishing individual violent acts, and the general procedures under which the content analysis was to be conducted. Feedback from the coders themselves was used in the early stages to confirm that the operational procedures of this exercise were clearly understood.

Intercoder consistency was checked both qualitatively and quantitatively. Qualitative checks comprised continuous monitoring of the judgments made by coders at key decision points in the application of the coding frame. Returned coding sheets were checked thoroughly for inaccuracies. Consistent errors triggered additional discussion of a coding decision among coders and researchers that were designed to resolve specific coding decisions.

Quantitative checks on reliability of coding focused on levels of agreement between pairs of coders in their identification of violent acts within particular programs and levels of consistency in the application of attribute codes and assignment of code values to individual violent acts. Reliability checks at the program level involved double coding of a quasi-randomly selected 10% subsample of the total program sample, with the limiting factor being the equal representation of programs from different channels. There were 15 cases (i.e., programs) of double coding in 1997 and 37 cases of double coding in total. In respect of violent acts and violence minutage measures, reliability coefficients were calculated using the alpha coefficient outlined by Krippendorf (1980). There is no simple solution for the problem of deciding the proper level of reliability. It is important to ensure not only that key measures of content achieve reliability but also remain relevant and meaningful in the context of the investigation (Holsti, 1969). In both years, reliability coefficients achieved respectable levels for violent acts (.94 and .92) and violence minutage (.92 and .91).

Attribute coding reliability was measured by having all coders separately classify the characteristics of 10 individual prerecorded violent acts selected from major program genres typical of those transmitted to the island. A coder by attribute matrix was created for each violent act and the data entered into the Holsti formula. The average level of intercoder agreement across attribute codes ranged between 58% (historical setting) to 89% (type of injury) over the 2 years. The overall average level of agreement was 79%. Only those attributes scoring above 75% are discussed in this chapter. The lowest reliability score was for historical context and resulted from occasional difficulties experienced by coders in being able to precisely place a scene in terms of its time period. For example, it was not always clear whether a program or a scene within a program was pre- or post-1980.

Amount of Violence on St. Helena Television

There were two principal measures of the quantity or amount of violence on television: (a) the total number of violent acts and (b) the total amount of transmission time occupied by violent material. There were also three principal measures of prevalence of violence: (a) the proportion of programs that contained any violence, (b) the average number of violent acts per violence-containing program, and (c) the proportion of total program running time occupied by violence. Results were compared across the 2 years of analysis.

A total of 679 violent acts were counted in both 1997 and 1998 (see Table 3.2). In all, 81 programs (59.6%) contained any violence in 1998, an increase on the previous year's figure of 71 (46.4%). These acts produced an aggregated total of 79.5 min of violence in 1998, compared with 41.7 min in 1997 and represent a year-on-year increase of 37.8 min of violent content. Thus, violence occupied 1.6% of total monitored program running time in 1998, an increase on the year before (1%). Much of the violence consisted of very brief incidents. The great majority of violent acts were just 1–2 sec (62% in 1998 and 64.6% in 1997) or 3–5 sec duration (15.4% in 1998 and 22.3% in 1997). More than one in ten (12.3%) violent acts in 1998 were in excess of 10 sec in duration compared to one in 20 (5.3%) in 1997. Violence-containing programs contained an average of 8.4 acts and 0.98 min of violence per program in 1998 compared with 9.6 acts and 0.59 min the year before. Thus, although the average rate of violence showed a marginal year-on-year decrease, the average amount of violence per violence-containing program increased by 71%.

Tables 3.3 and 3.4 show the distribution of violence by program genre. The first of these two tables shows the distribution of violent acts and the second shows the distribution of minutes of program running time occupied by violence in 1997 and 1998. Children's cartoons contained the

TABLE 3.2
Amount of Violence on Broadcast Television

	1997	1998	Change
No. progs with any violence	71	81	+10
Percent progs with any violence	46.4	59.6	+13.2
Total no. violent acts	679	679	0
Amount of violence (mins)	41.7	79.5	+37.8
Ave. No. acts per viol. prog.	9.6	8.4	−1.2
Ave. mins viol.per viol.prog.	0.59	0.98	+0.39

TABLE 3.3
Distribution of Violent Acts by Program Genre

	1997		1998	
	Number of Violent Acts	Percent of Violent Acts	Number of Violent Acts	Percent of Violent Acts
Cinema films	53	7.8	28	4.1
UK drama	96	14.1	—	—
US drama	15	2.2	79	11.6
Other drama	8	1.2	26	3.8
Cartoons	170	25.0	315	46.5
Sport- Live	230	33.9	43	6.3
Sport highlights	13	1.9	66	9.7
News	84	12.4	81	11.9
Current affairs	2	0.3	29	4.3
Documentary	8	1.2	8	1.2
Sitcoms/Arts	—	—	3	0.4
Total	679	100.0	679	100.0

TABLE 3.4
Distribution of Violence in Minutes by Program Genre

	1997		1998	
	Amount of Violence in Minutes	Percent of Violence	Amount of Violence in Minutes	Percent of Violence
Cinema films	3.5	8.4	4.9	6.1
UK drama	8.2	19.8	—	—
US drama	0.6	1.3	6.4	8.0
Other drama	0.8	1.8	1.0	1.3
Cartoons	7.7	18.4	15.6	19.6
Sport- Live	19.2	46.3	40.0	50.2
Sport highlights	0.7	1.6	3.9	4.9
News	0.4	1.0	4.6	5.8
Current affairs	0.1	0.1	2.7	3.4
Documentary	0.5	1.3	0.5	0.6
Sitcom/arts	—	—	0.1	0.1
Total	41.7	100.0	79.5	100.0

greatest number of separate violent acts of any genre followed by news programs and US-produced drama. These three genres together accounted for 7 out of 10 violent acts.

When looking at the distribution of violence in terms of time occupied by violent acts, live sports emerged as the single most significant contributor of any genre. Around half of all violence occurred in these programs with most of the remainder occurring in children's cartoons. Year on year there were significant increases in the proportion of violent acts contributed by cartoons ($\chi^2 = 6.5$, $df = 1$, $p < .02$) and US drama programs ($\chi^2 = 6.4$, $df = 1$, $p < .02$). Furthermore, whereas UK drama programs had contributed 1 in 7 violent acts in 1997, no violence at all was contributed by this genre in the period monitored in 1998. The proportion of violent acts contributed by live sports broadcasts showed a significant decrease from one year to the next ($\chi^2 = 18.9$, $df = 1$, $p < .001$).

There were some shifts in the level of contribution different genres made to minutes of violence across the two periods of analysis. Year after year, however, the amount of violence registered in sports broadcasts more than doubled from 19.2 min in 1997 to 40 min in 1998. As a proportion of total program running time, the greatest amount of violence occurred in children's cartoons where violence occupied 4.0% of running time, followed by sport (2.6%), drama (1.4%), and, finally, factual programs (0.4%). The proportion of violence (in minutes) contributed by US drama programs increased significantly from 1997 to 1998 ($\chi^2 = 4.8$, $df = 1$, $p < .05$), whereas that contributed by news programs increased by a degree that almost achieved statistical significance ($\chi^2 = 3.4$, $df = 1$, $p < .06$). The proportions of violence (in minutes) contributed by other genres did not change significantly.

Despite the significant increase in overall amount of violence on broadcast television in St. Helena, year on year, perspective is added by noting that in both years the 10 most violent programs accounted for a substantial proportion of the total amount of violence catalogued overall. In 1997, 60.9% of all violent acts were accounted for by the 10 most violent programs, whereas in 1998, the top 10 contained 39.6% of all violent acts. In 1997, 81.9% of all violence in terms of minutes of program running time occurred in just 10 programs, whereas in 1998 the equivalent figure was 70.5%. Boxing and wrestling bouts accounted for significant amounts of all violence. In 1997, two televised boxing matches and one wrestling tournament accounted for 42.9% of all violence minutes and in 1998, 47% of all the minutes of violence registered was contributed by just two broadcasts—a World Wrestling Federation tournament and a heavyweight championship boxing match.

Attributes of Violence

Violent acts were classified in terms of their physical form, the contexts and settings in which they occurred, their consequences for those involved, and the types of characters who featured as either aggressors or victims. The physical form of violence was categorized according to 29 types of behavior and 31 categories of weapon or instruments of violence.

The most frequently occurring forms of violence were similar across the 2 years of program monitoring. The most frequently coded forms of violence on broadcast television in St. Helena were incidents involving the perpetrator physically pushing, tripping, or punching a victim. Otherwise, aggressors tended to kick, shoot, club, or verbally abuse their targets. The main changes, year on year, were in the rankings of these violent acts (see Table 3.5). Year on year, however, there were no significant changes in proportions of violent acts that involved pushing and tripping ($\chi^2 = 1.93$), punching ($\chi^2 = 1.92$), verbal abuse ($\chi^2 = 1.74$), or shooting ($\chi^2 < 1.0$).

A further analysis was conducted to reveal the 10 most frequently occurring types of instruments of aggression (see Table 3.6). By far the most frequently used instrument of aggression was the hand or fist. Otherwise, perpetrators of violence used some other body part. In 1998, a new category of "voice" was included and showed up prominently, further reflecting the prevalence of verbal aggression in the form of violence analysis. In contrast to the latter analysis, however, there were more year-on-year shifts in the composition of the top 10 for types of weapon. Whereas stones, whips, swords, and knives featured prominently in 1997, by 1998 these had been replaced by rifles and shotguns, other stabbing instruments, throwing instruments, and driven vehicles. The proportions of violent acts that involved the fist or hand ($\chi^2 < 1.0$), foot ($\chi^2 = 1.1$), other body part ($\chi^2 < 1.0$), or pistol/handgun ($\chi^2 < 1.0$) did not change significantly across the 2 years.

The most violent program genres in purely quantitative terms were live sport and children's cartoons. This pattern occurred across both years of analysis. It is important though to look beyond simple frequency counts at the forms of violence typically displayed in different genres. Most of the sports violence occurred in televised wrestling bouts and included a vari-

TABLE 3.5
Most Frequently Occurring Types of Violence

Violence Type 1998	No. of Violent Acts	% of Violent Acts	Violence Type 1997	No. of Violent Acts	% of Violent Acts
Push/trip	109	16.1	Push/trip	170	25.0
Punch	76	10.2	Punch	119	17.5
Verbal abuse	59	8.7	Kick	54	8.0
Shoot	53	7.8	Shoot	48	7.1
Hit with club	36	5.3	Hit with club	36	5.3
Kick	35	5.2	Verbal abuse	27	4.0
Property damage	33	4.9	Slap	21	3.1
Slap	26	3.8	Bomb explosion	17	2.5
Military bombardment	15	2.2	Property damage	14	2.1
Bite/scratch	13	1.9	Military bombardment	12	1.8

TABLE 3.6
Most Frequently Occurring Types of Weapon

Weapon Type 1998	No. of Violent Acts	% of Violent Acts	Weapon Type 1997	No. of Violent Acts	% of Violent Acts
Fist/hand	270	39.8	Fist/hand	266	39.3
Voice	48	7.1	Foot	73	10.8
Foot	44	6.5	Other body part	63	9.3
Other body part	41	6.0	Bow & Arrow	26	3.8
Rifle/shotgun	27	4.0	Stone	17	2.5
Club	23	3.4	Pistol/handgun	16	2.4
Pistol/handgun	21	3.1	Club	16	2.4
Stabbing instrument	16	2.4	Whip/cane	12	1.8
Throwing instrument	15	2.2	Sword	10	1.5
Vehicle	11	1.6	Knife	10	1.5

ety of forms of violence involving the use of the body, usually taking the form of a punch, a push or trip, or a kick. Verbal abuse was common in wrestling. In boxing, understandably, most of the violence comprised punching. In cartoons, for example, there were many varied and sometimes unusual forms of violence with the most frequent being pushing, hitting with a club, verbal threats and abuse, property damage, punching, stabbing, biting or scratching, and slapping. Children's cartoons was the only program genre to depict stabbing scenes and the only genre with violence classified as bullying or cruelty, hanging scenes, explosions, fire setting, and electrocution. There were similarities with the forms of violence recorded in children's programs in Britain except that in the latter television market, shootings featured as the most frequently occurring type of violence of all in children's programs (Gunter & Harrison, 1997).

In more serious drama programming aimed at adult audiences, the principal forms of violence were verbal abuse, pushing, property damage, slapping, and punching. Small numbers of scenes (fewer than four in each case) featured kicks, biting or scratching, use of a spraying device, and strangulation. In cinema films, violence was restricted to shooting, pushing, verbal abuse, kicking, and slapping.

Motivational Context of Violence

The context of violence was classified according to eight categories that qualified violence in terms of the dramatic motives that appeared to underpin it. Across 1997 and 1998, these included civil strife (13.4% and 9.3%), interpersonal conflicts or arguments (12.9% and 36.1%), upholding the law (4.6% and 3.7%), self-defense (2.7% and 1.6%), committing a crime (2.4% and 2.4%), war or military combat (2.2% and 2.2%), or domestic violence taking place in the home (2.2% and 19.5%). Other violent acts were not unambiguously classifiable under these headings (58.9% and 17.2%). The major shifts occurred in regard to interpersonal arguments ($\chi^2 = 11.0$, $df = 1$, $p < .001$) and domestic violence ($\chi^2 = 13.8$, $df = 1$, $p < .001$), both of which increased year on year by significant margins.

Another measure of the motivational context of violence was classification of the goals of aggressors. Violence was distributed across nine separate types of aggressor goal. Four of these goal types were altruistic or defensive: upholding the law (6.6% in 1997 and 10.9% in 1998 of violent acts), self-preservation (5.9% and 10.1%), protection of society (0.9% and 8.4%), and protection of the family (3.7% and 8.3%). Five further goals were more selfish or antisocial: acquisition of money (3.1% and 8.7%), ambition or power (2.2% and 6.5%), evil and destructive reasons (4.4% and 6.6%), sexual motives (0.4% and 0.1%), and religious motives (1.0%). Thus, violence in the name of selfless or altruistic motives underpinned a larger proportion of violent incidents than did violence for more selfish or suspicious motives (17.1% and 37.7% vs. 10.1% and 22.4%). Only the proportion of violent acts motivationally attributed to the protection of society, however, exhibited a significant shift year on year ($\chi^2 = 6.1$, $df = 1$, $p < .02$).

Consequences of Violence

Violence was also classified in terms of the degree of injury incurred by victims. More than half of all violent acts produced no injurious consequences that could be readily seen on screen (61%, an increase on 52% in 1997). Minor injuries resulted from one in four acts in 1997 and from one in eight acts in 1998. The latter shift was statistically significant ($\chi^2 = 4.2$, $df = 1$, $p < .05$). Serious injury and death were relatively rare consequences of violence across both years, with the occurrence of fatalities showing a drop (nonsignificant), year on year (see Table 3.7).

TABLE 3.7
Degree of Injury Caused

	1997 %	1998 %
No injury	52.4	61.0
Minor injury	25.3	12.7
Serious injury	1.6	1.3
Death	3.8	0.9
Psychological damage	0.3	3.8
Unclear	16.6	20.3

Aggressors and Victims

One of the key factors of interest in any investigation of the representation of violence on television is the gender of the aggressors and victims. Table 3.8 shows the distribution of male and female aggressors and victims across both years. The great majority of on-screen aggressors and victims of violence were male. There was a significant increase in the number of females involved in violence as aggressors on St. Helena television across the two periods of monitoring ($\chi^2 = 7.1$, $df = 1$, $p < .01$). Although there was an increase in the proportion of female victims of violence ($\chi^2 = 1.8$, ns) and a concomitant decrease in the proportion of victims who were male ($\chi^2 = 2.5$, ns), across the two monitoring periods, in neither case were these changes statistically significant.

The age distribution of aggressors and victims was classified according to nine categories: babies and toddlers, preschool children, young children (5–11), children (12–14), teenagers (15–19), young adults (20–35), middle-aged adults (36–64), 65+, and mixed age groups. The largest proportion of aggressors fell into the young adult (39.8% in 1998 and 46.8% in 1997) and middle-aged adult (7.2% in 1998 and 16.7% in 1997) age groups, as did the largest proportion of victims of violence (young adults: 37.8% in 1998 vs. 46.6% in 1997 and middle-aged adults: 5.4% in 1998 vs. 11% in 1997). In the 1998 analysis, there were 47 violent acts in which children up to the age of 11 featured as aggressors and 23 incidents in which they were victims of violence. Young teens aged 12–14 were more likely than older teens (15–19) to be aggressors (28 vs. 10 incidents) or victims of violence (26 vs. 17 incidents). There were still many incidents in which the age of aggressors (35.2%) and victims (42.6%) was unclassifiable. In 1997, just 22 incidents were coded in which children and teenagers were featured as aggressors and 54 incidents in which they were victims. The great majority of child and teenage victims of violence (41 of 54)

TABLE 3.8
Aggressor/Victim Gender

	1997 %	1998 %
Aggressor		
Male	84.6	78.8
Female	5.3	18.2
Other/unclear	10.1	3.0
Victim		
Male	81.2	62.2
Female	6.9	12.9
Other/unclear	11.9	24.9

were in their late teens (15–19 years). Hence, there was a year on year increase in involvement of younger children in violence on broadcast television in St. Helena. Across the 2 years, the proportion of aggressors who were children or teenagers (χ^2 = 5.5, df = 1, p < .02) increased significantly, whereas the proportion of victims from the age group (χ^2 < 1.0) did not alter significantly.

Finally, the dramatic roles occupied by aggressors and victims were examined in closer detail. For this analysis, aggressors and victims were classified in 31 categories of role types. These categories represented different occupational groups and specific dramatic categories. The results for aggressors and victims of violence are summarized separately in Tables 3.9 and 3.10.

Among both aggressors and victims, the predominant categories were cartoon characters and sports persons. This finding is not surprising given earlier results showing that cartoons and sports broadcasts (i.e., boxing and wrestling) contained more violent incidents than other genres. Members of the public, criminals, and police were three other role categories that featured within the most frequent types of aggressor and victim across both years. The nonspecific category of "general occupation" also featured among the most frequently occurring aggressor and victim role types during both years.

Year on year, there were significant increases in the proportions of violent acts perpetrated by human cartoon characters (χ^2 = 6.4, df = 1, p < .02) and by members of the public (χ^2 = 5.6, df = 1, p < .02), and a significant decrease in violent acts perpetrated by sports persons (χ^2 = 5.9, df = 1, p < .02). None of the other top 10 character types that featured both in

TABLE 3.9
Most Frequent Types of Aggressor

	1998			1997		
Type of Aggressor	No. of Violent Acts	% of Violent Acts	Type of Aggressor	No. of Violent Acts	% of Violent Acts
Cartoon animal	145	21.4	Sports person	246	36.3
Cartoon human	135	19.9	Cartoon animal	111	16.3
Sports person	124	18.3	Cartoon human	46	6.8
Member of public	95	14.0	Knight	42	6.2
Armed forces	59	8.7	Period actor	40	5.9
General occupation	12	1.7	Police uniformed	28	4.1
Rioter	12	1.7	Member of public	27	4.0
Criminal	10	1.4	Hooligan	23	3.4
Police uniformed	8	1.2	Criminal	15	2.2
Cartoon other	7	1.0	General occupation	14	2.1

TABLE 3.10
Most Frequent Types of Victims of Violence

| | 1998 | | | 1997 | | |
Type of Victim	No. of Violent Acts	% of Violent Acts	Type of Victim	No. of Violent Acts	% of Violent Acts
Cartoon animal	145	21.4	Sports person	246	36.5
Sports person	123	18.1	Cartoon animal	102	15.0
Cartoon human	120	17.7	Cartoon human	62	9.2
Member of public	83	12.2	Member of public	54	8.0
Armed forces	21	3.1	Period actor	48	4.1
Rioter	12	1.8	Knight	28	4.1
Criminal	10	1.5	Police uniformed	22	3.2
Police—plain clothes	10	1.5	Criminal	14	2.1
General occupation	10	1.5	General occupation	11	1.6
Cartoon other	5	0.7	Armed forces	11	1.6

1997 and 1998 exhibited significant shifts in prevalence. Among the victims of violence, only sports persons exhibited a significant change in frequency of occurrence across the two monitoring periods ($\chi^2 = 6.2$, $df = 1$, $p < .02$).

DISCUSSION AND IMPLICATIONS

This chapter examined the nature of broadcast television output in St. Helena with special reference to the representation of violence in programs. The study was based on modest samples of television output on the island's two channels recorded in 2 separate years. Findings revealed that violence was a prevalent feature of television output, though probably does not qualify as a prominent feature of programs transmitted.

In 1998, a larger proportion of programs monitored (59.6%) contained any violence when compared with programs in 1997 (46.4%). Although the program sample in 1998 ($n = 136$) was smaller than that monitored in 1997 ($n = 153$), exactly the same number of violent acts was cataloged over both years (679), but in 1998 this violence occupied a greater total amount of program running time than had been found in 1997 (79.5 min vs. 41.7 min). Even so, three quarters of all violent incidents lasted less than 5 sec and 70% of this violence was found in just 10 programs or 7% of the programs monitored. In fact more than half the minutes of violence occurred in just two programs. Hence, although the quantity of violence on St. Helena television increased year on year, this finding must be put in perspective by examining its distribution across the output.

The findings for St. Helena can be benchmarked against the violence observed on broadcast television in Britain in the 1994–1996 period (see Gunter & Harrison, 1997a, 1997b, 1998). Average rates of occurrence of violent acts in violent programs and average amounts of violence in minutes per violent program were found in 1997 to be lower than in the UK (9.6 acts vs. 10 acts; .59 min vs. 1.55 min). In 1998, the average rate of violence per violent program on broadcast television in St. Helena fell to 8.4 but the average amount of violence in minutes increased to .98 min, still remaining below the UK average figure. However, when the violence contributed by the two most violent programs (a wrestling bout and a boxing match) was subtracted from the total amount recorded in 1998, the average amount of violence in minutes for the remaining violent programs was .53 min. The latter figure represented a decrease on the already low 1997 figure for St. Helena broadcast television.

Concerns about children's exposure to violence stem both from the overall amount of violence to which they can be exposed and the nature of that violence. Approaching half of all the violent acts cataloged in this study occurred in cartoon programs aimed at the child audience, and as these occurred in just 3 days of transmissions, it is clear that this genre represents the form in which most violence is likely to be seen by children on the island. Whether this can be regarded as a finding that merits serious concern depends upon how problematic cartoon violence is perceived to be. One significant factor is whether children themselves take cartoons seriously enough to copy the behavior they see portrayed in such programs or to learn longer-term lessons in social behavior from watching animated characters. Some research evidence has indicated that children draw distinctions between the highly fantastic contexts of cartoons and reality (Hodge & Tripp, 1986). Children's viewing tastes and diets change, however, as they grow older and display increasing interests in more serious drama productions and factual programming.

Some cartoons have been identified as deserving of special attention because of their popularity, the fact that the programs are often closely connected with wider merchandising campaigns, and because anecdotal observations have occasionally suggested that youngsters appear to copy the behavior displayed in these programs. Such concerns have focused most especially on certain children's action-drama series such as *The Teenage Mutant Ninja Turtles* and *The Mighty Morphin' Power Rangers*, in which human or humanlike characters display a frequent use of violence and an apparent invulnerability to harm (see Boyatzis, Matillo, & Nesbitt, 1995). Such programs were not broadcast to St. Helena children during the periods of either of the 2 years of television content analysis.

From the perspective of scheduling, it is clear that there were plentiful opportunities for children to witness screen violence on the 6 days of television transmissions monitored in this study. Besides the cartoons that were aimed at young audiences, and were shown during daytime, the wrestling and boxing were televised at times when many children, potentially, could have been watching. The most violent television drama, "Assignment Berlin," however, was not shown until 10:00 p.m. on Sunday evening. Implications about the significance of televised violence for children as indicated by a study such as this, can primarily be found in its analysis of the forms of violence that occur, the contexts and settings in which they occur, how much harm or damage is caused, and the types of people or characters who feature as perpetrators or victims of violence.

The nature of the violence shown on broadcast television in St. Helena displayed a largely similar character from one year to the next. The most commonly occurring types of violence coded in this study comprised largely mild physical assaults where one character pushed, shoved, slapped, or hit another. There were incidents of shooting and other episodes in which bombs and military equipment were used. The latter forms of violence were mostly found in news broadcasts, however, and took place in geographically distant locations. There were few incidents that featured more severe forms of violence, such as stabbing or strangulation, that are known to upset viewers (Gunter, 1985). Violence that involved the use of sharp instruments almost inevitably took place in animated cartoon contexts in which other attributes associated with serious audience reactions, such as the infliction of serious injury, pain, and suffering in settings close to home were absent (see Potter et al., 1996; Wilson et al., 1996).

Turning to the motivational context of the violence, the results were similar across the 2 years. Much of the violence cataloged in this analysis was apparently motiveless. This finding reflected the major contribution made to overall violence levels by live sports broadcasts and cartoons, where perpetrators' specific motives for violence may not be immediately apparent. For those incidents which could be categorized by motivational context or goal of aggressor, however, violence in service of good generally outweighed violence in the service of evil or negative motives.

A majority of all the violence coded produced no observable harm to victims (52% in 1997 and 61% in 1998). When harm did occur, it tended to take the form of only minor injuries. Once again, this pattern is consistent with observations made of UK television, on which 42% of violent acts caused no apparent injury or harm to victims. When injuries did occur, they tended to be minor rather than serious (Gunter & Harrison, 1997a, 1998). The results here reflect the significant contributions of American professional wrestling and cartoons to overall violence levels. In neither of these two genres does violence tend to result in injury or harm to those on the receiving end.

Past observations of children at play have indicated an increase in the adoption of super-hero themes (French, Pena, & Holmes, 1987; Jennings

& Gillis-Olion, 1980; Kostelnik, Whiren, & Stein, 1986). Television programs featuring superhero cartoons have been observed to have increased significantly following marketing and promotional developments between the toy industry and broadcast organizations from the mid-1970s. Concern has grown that such programs, along with the reinforcement of toy action figures, may be contributing to the increase of children's adoption of superhero themes in their imaginative play behavior (French & Pena, 1991).

In the current research, perpetrators of violence were dominated by sports stars and cartoon characters. The contexts of sports violence were contained within specific competitive environments. Such violence should be distinguished from violence that occurs in the context of a storyline in which an attractive character and potential role model for children is depicted using violence to overcome a problem, and through which certain social values might be transmitted to young viewers so that the use of violence under certain circumstances becomes acceptable. It was clear from the age profile of perpetrators of violence that very few fell in child and teenage age groups, with whom young viewers might be expected most closely to identify (Maccoby & Wilson, 1957; Reeves, 1979). Although it should perhaps be added that the idea of "identification" is a complex concept and there may be a world of difference between a child saying that they like a particular television character, would want to be like that character, and then actually trying to emulate that character's behavior (Buckingham, 1993).

Violence shown on television in St. Helena, as with many other countries, was a predominantly male behavior. There were relatively few female perpetrators of violence as well as relatively few female victims. This pattern was significantly shaped by the dominance of sports-related violence, which occurred without exception between male contenders. Two interesting trends of significance were the increase in female involvement in violence across the two monitoring periods, especially as aggressors, and the increase in numbers of child and teenage aggressors. Although these groups represented only small minorities of those involved in violence on television, there was a distinct upward year-on-year trend in which females and youngsters, especially preteens, got involved in violent behavior.

In conclusion, the analysis of the representation of violence on St. Helena television revealed patterns of portrayals not dissimilar to those found on mainstream television systems in other countries (Gunter & Harrison, 1998; McCann & Sheehan, 1985; Watson et al., 1991; Wilson et al., 1996). Average rates of occurrence of violent portrayals on St. Helena television were found to match those of other communities in which television has been long established. Once again, simple frequency counts were found to give a misleading impression of the prevalence of televised violence. As observed elsewhere (Gunter & Harrison, 1997a, 1998), a few programs can disproportionately account for the total amount of violence on television.

Comparing the findings across the two periods of analysis, there was an increase in the total amount of violence detected, despite the fact that similar quantities of program output were analysed on both occasions. The concern that such a finding might generate on the surface is reduced when noting that this increase was mostly accounted for by a small number of live sports broadcasts featuring boxing and wrestling. Further, the nature of violence depicted on St. Helena television did not exhibit many changes of character across the two periods of analysis. Where such changes did occur, however, they indicated increases in domestic violence and violence arising out of arguments between two or more people, a decrease in violence causing only minor injury, and an increased involvement of females and young people in violence as aggressors. These findings are singled out because they achieved statistical significance. The minor injury finding is perhaps of least social significance because there were counterbalancing (though nonsignificant) decreases in violence causing serious injury or fatalities. Increased domestic violence can be regarded as a "risk factor," because it represents greater potential realism (Wilson et al., 1996). Meanwhile, the greater involvement of females and young people in acts of aggression where they are the perpetrators could have implications for potential modelling effects among young people in the audience. Results, presented later in this book, regarding children's social behavior indicate that although no evidence has emerged yet to indicate major shifts in behavior patterns, young children (especially boys) who exhibited antisocial tendencies before the onset of broadcast television later showed preferences for viewing programs with violence. Such findings indicate the need for continued monitoring of television output on St. Helena.

CHAPTER FOUR

The Impact of Television
on Children's Leisure

Andrew Hannan
University of Plymouth, UK

It is highly unusual to find anything resembling a "natural experiment" where researchers are provided with an opportunity to observe the effects of factors held to be influential but impossible to isolate. One such factor is television. All know it is supposed to have effects on children's behavior (Gunter & McAleer, 1990), but it is very difficult to measure these (Gauntlett, 1995; Hodge & Tripp, 1986), especially in natural rather than artificially contrived circumstances. A number of studies including Brown et al. (1974) in Scotland, Murray & Kippax (1978) in Australia, Williams & Handford (1986) in Canada, and Mutz et al. (1993) in South Africa, have attempted this by studying the impact of television in communities with different levels of exposure or by monitoring responses to its introduction. One of the earliest and most influential studies was undertaken in the UK by Himmelweit et al. (1958), who compared children who were viewers with controls who were not, making use of a survey of children's leisure habits by means of diaries kept by children aged 10–11 years and 13–14 years. They also undertook a rare "before and after" study of the introduction of television to Norwich. The research reported here is a somewhat similar study that took place over the period 1994–1998 of the effects of the introduction of broadcast television in St. Helena on children aged 9–12 years.

A series of surveys based on children's diary entries have been carried out on St. Helena that have attempted to record children's "leisure" behavior and other activities undertaken outside school including "work" in or outside the home and daily tasks such as washing and eating. The first diary–survey took place in October 1994 (before television), the second at the end of November and beginning of December 1995, approximately eight months after the introduction of Cable News Network broadcasts, the third in February 1997, approximately 4 months after the addition of M-Net, when viewers were offered a mix of CNN, Hallmark (films), Discovery Channel, Supersport and Cartoon Network (M-Net programs were available as well as CNN for an extra payment), and the fourth in October 1998. In these last two stages, viewing was possible around the clock with no equivalent of the 9 p.m. threshold for "adult" programs that operates for terrestrial television in the UK (for information on the content of broadcasts see Gunter et al., 1998). This diary-survey method has made it possible to monitor the changing patterns of children's self-reported leisure activities and to assess the impact of the introduction of television. We have here, then, some of the characteristics of a "natural experiment," although, as we shall see, the real world keeps complicating matters and spoiling the purity of the research "design."

OTHER STUDIES

Direct comparisons of St. Helena with other locations in terms of information about children's leisure activities are not easy to make. The report of the study by Himmelweit et al. (1958) does not give separate detailed information about the results of the diary-based survey and in any case is very dated with the information being collected in 1955, forty years before this study. In general, their study found that the introduction of television was followed by a reduction in radio listening, comic reading, and cinema attendance. They explained this in terms of a theory of "functional similarity" (i.e., that television replaces those activities where it better satisfies certain needs that have previously been met by the most displaced activities). Himmelweit et al. also put forward the proposition that television tends to reduce the amount of time spent in relatively unorganized "marginal" activities. However, they found that the initial loss of time suffered by some activities, notably book reading, was not prolonged, suggesting that there was a "novelty" effect that wore off for those activities that were sufficiently different from watching television.

Another important study of the displacement effects on children's leisure activities was undertaken in Scotland by Brown et al. (1974), who compared similar small isolated rural communities, one of which, Arisaig, was going through the process of receiving television for the first time. The major contribution of this paper is its critique of the "functional similarity" theory. Brown et al. argue that a theory of "functional reorganization" better explains the complex of factors at work, stating that "Old media never die; they only change their functions" (p. 110). They found that the introduction of television to Arisaig resulted in a decline of all other categories of leisure time activity. The average number of reported "indoor" leisure activities undertaken at least weekly (excluding watching television) fell from 5.5 to 3.83. Those that were "outdoor rule governed" (e.g., soccer, hide-and-seek, and country dancing) fell from 4.44 to 2.94 and the "outdoor nonrule governed" (e.g., cycling, fishing, and playing on swings) from 4.06 to 3. However, the numbers involved in the empirical study were very small (just 18 children in the village gaining television) and the methods of recording leisure activities were fairly imprecise (asking by interview whether media such as listening to the radio or reading a book had been accessed "every day," "most days," "once or twice a week," or "hardly ever").

The study by Murray & Kippax (1978) that was based on interviews with a total of 128 children aged 8–12 years in three towns in Australia, is of particular interest as they discovered that the effects of the introduction of television were not simply displacement. They found that:

> In some instances, such as involvement in outdoor activities and watching and playing sport, there were linear decreases associated with increasing availability of television. On the other hand, children's involvement in playing with friends and toys and sitting around doing nothing manifested linear increases with increasing availability of television. (p. 27)

Like Himmelweit et al., they discovered that those children who had long been exposed to television tended to return to their former (pre-TV) levels of interest in other media and social participation activities, although pursuits such as watching sport, outdoor activities, and reading comics (but not books or newspapers), which fell sharply initially, tended not to recover in this manner. Such findings, however, were based on comparisons of different communities with various levels of exposure to television rather than a longitudinal study of increasing exposure to television in a single community.

A longitudinal study was included in the research undertaken by Williams & Handford (1986) who reported the pattern of leisure activities in "Notel," a town in Canada, both before and after the availability of television. Their investigation was based on a questionnaire survey that asked residents (adults as well as high-school students aged 13–18 years) which activities they had engaged in during the last year and how frequently (ranging from once a year to daily). They concluded:

> Whereas the effect of television on participation in community activities, especially active participation in sports, was substantial, its impact on partici-

pation in private leisure activities, as assessed by either frequency or number, was minimal. (p. 184)

However, it is difficult to discover from these findings just what the precise level of each activity was in Notel before the introduction of television. In any case, the frequency was given in terms of participation over
the course of the previous year rather than a detailed account of activities
engaged in for each hour of a particular day or days. Furthermore, the age
groups used were outside the scope of the present survey.

More directly relevant research is reported by Mutz et al. (1993) who
describe the large-scale longitudinal project undertaken among White
schoolchildren in South Africa over a period of 8 years, beginning in 1974,
2 years before the introduction of television. This made use of a panel of
approximately 1,900 students who, beginning as 11-year-olds, took part in
each of the annual data collections. It also employed cross-sectional samples of 1,500 students at each of the eight grade levels (from 5th to 12th
grade). Mutz et al. conclude that the magnitude of television's influence
on what they call "important activities" (watching movies, listening to the
radio, reading, taking part in sports activity, doing homework, attending
clubs or out-of-school lessons, engaging in hobbies) "is modest at best" (p.
70). They found that, "When television became available, over 60% of
youngsters' viewing time came from marginal activities" (p. 71). Unfortunately, they do not attempt to discover the nature or measure the extent of
these "marginal activities," possibly being restricted by the content of the
questionnaire surveys they carried out, which do not seem to have asked
the students to present diary accounts of all the time they spent outside of
their school activities. Nonetheless, the paper is notable for its excellent
critique of displacement theory and, in particular, for showing that reductions in the extent of television watching did not result in the restoration
of activities that had lost ground to their previous levels of eminence (a
finding which conforms to the functional reorganization rather than the
functional similarity theory).

Clearly, though, the monitoring of the impact of television's availability
upon youngsters' leisure time pursuits has received inadequate attention.
It can be argued, also, that television watching is itself an activity that has
changed over time, given the shifts in the nature and extent of program
provision as well as program content since the 1950s. More recent and detailed information on children's leisure activities before and after the availability of television is needed to enhance our knowledge and understanding of the ways in which (and the extent to which) watching television
impinges on aspects of youngsters' social lives. The St. Helena Research
Project offered an almost unique opportunity to undertake such inquiry in
an English speaking, westernized culture.

THE POPULATION

It was intended to include in the survey all 9- to 12-year-old children on the island by targeting those attending all three of the island's middle schools (St. Paul's, Pilling, and Harford). In 1994 a total of 269 pupils took part, which represents all those present at the schools on the days involved (except for a missing batch of Year 3 pupils from one of the schools). This number fell to 258 in 1995, to 206 in 1997, and to 167 in 1998, as any of those for whom the coding of identities was inconsistent were excluded from the analysis (so as to be sure that children given the same code number were indeed the same individuals), as well as losing those included previously who were absent on subsequent occasions. The characteristics of the respondent population are given in Table 4.1.

METHOD

The research was based on the use of a diary form from Himmelweit et al. (1958). See Fig. 4.1 for the version used for a weekday.

In the first (pre-TV) phase in 1994, the diaries were kept for Sunday, October 9, Tuesday, October 11, and Thursday, October 13 (other demands on both teachers and pupils made it impracticable to obtain records of a whole week). In the second (post-CNN) phase in 1995 the schools were unable to use the same days: St. Paul's used Sunday, Tuesday, and Thursday, November 19, 21, and 23; Harford used Sunday, Tuesday, and Thursday, November 26, 28, and 30; Pilling used Thursday, November 30, Sunday and Tuesday, December 3 and 5. In 1997 in the third (post-M-Net) phase, all three schools used Sunday, Tuesday, and Thursday, February 16, 18, and 20. In the fourth phase, in 1998, the days used were Sunday, Tuesday, and Thursday, October 11, 13, and 15. Ideally, of course, all schools would have used the same week at the same time of year in each phase but this was not possible to arrange. However, although some allowance needs to be made for the different times of year (e.g., the high number of time slots devoted to "indoor hobbies" in 1995 when children were making various items for Christmas celebrations), seasonal variations in temperature and hours of daylight are not great.

The class teachers introduced the exercise to the children using the same form of words set out in Himmelweit et al. (1958, pp. 417–418), with similar respondent coding devices being employed and teachers being asked to provide information on each child in terms of year group and gender (see Table 4.1). The variations from the Himmelweit et al. original (see Figure 4.1) were confined to substituting 'After 10 o'clock in the evening' for 'Between 10 and 11 o'clock in the evening' and using 'Before 8 o'clock in the morning' as the first such time slot on the form for the

TABLE 4.1

Characteristics of Respondent Population for all Phases

	1994		1995		1997		1998	
	Number	*%*	*Number*	*%*	*Number*	*%*	*Number*	*%*
St Paul's school	124	46.1	122	47.3	99	48.1	73	43.7
Pilling school	77	28.6	62	24.0	59	28.6	59	35.3
Harford school	68	25.3	74	28.7	48	23.3	35	21.0
Year 2 (9–10 yrs)	101	37.5	91	35.3	44	21.4	52	31.0
Year 3 (10–11 yrs)	65	24.2	99	38.4	84	40.8	60	36.0
Year 4 (11–12 yrs)	103	38.3	68	26.4	78	37.9	55	33.0
Male	138	51.3	128	49.6	112	54.4	88	52.7
Female	118	43.9	118	45.7	94	45.6	79	47.3
Unknown	13	4.8	12	4.7				
Total	269		258		206		167	

To be filled in on_____the_____of_____	Code Number
DIARY SHEET FOR	
_____	_____
afternoon and evening	
WRITE DOWN ALL THE THINGS WHICH YOU DID	For each thing you did, say whether you did it by your-self or with someone else, say who this was
Between 4 and 5 o'clock in the afternoon	
Between 5 and 6 o'clock in the afternoon	
Between 6 and 7 o'clock in the evening	
Between 7 and 8 o'clock in the evening	
Between 8 and 9 o'clock in the evening	
Between 9 and 10 o'clock in the evening	
After 10 o'clock in the evening	

Please think back again over **everything** you did yesterday afternoon and evening. Write down the three things you **really** enjoyed:

a) I enjoyed most of all

b) I enjoyed next most

c) I enjoyed third most

FIG. 4.1. The diary form.

Sunday (which also asked for three ranked preferences across the whole day). Using these labels for the final slot on all 3 days and the first slot on the Sunday was intended to give children the opportunity to record activities undertaken unusually late and (in the case of Sunday) early in the day. However, there was also a complication here in that the forms for the second and third phases were erroneously amended to add an additional slot, so that the last two of these referred to 'Between 10 and 11 o'clock in the evening' and 'After 11 o'clock in the evening.' This, of course, complicates comparisons with the first and fourth phases. However, because the entries for 'After 11 o'clock in the evening' were almost always 'sleeping' (92% in phase two and 94% in phase three), it was decided not to include entries for this time slot in the analysis. Thus, for the second and third phases where children undertook activities other than sleeping after 10 p.m., these were almost always entered in the penultimate (10 p.m.–11 p.m.) rather than the ultimate (after 11 p.m.) slot. It was, therefore, justifiable to compare responses for the after 10 p.m. slot for 1994 and 1998 with those for the 10 p.m. to 11 p.m. slot for 1995 and 1997.

The coding and data entry were carried out making use of the categories set out in Himmelweit et al. (pp. 418–419), except for 'cinema visits,' 'reading newspapers,' and 'school,' which were not cited by the children in this survey, and adding others as appropriate. It should be noted here that distinguishing between entries for 'watching video' and 'watching TV' was a difficult task, especially in 1997 and 1998. However, these were recorded under the latter heading unless the prerecorded nature of the programs was clearly apparent. It must also be taken into account that, with the exception of 'Before 8 o'clock in the morning' for the Sunday diary and 'After 10 o'clock in the evening' for all 3 days, each of the slots was an hour in duration. However, the coder (who was the same person for all four phases) entered only one activity for each slot (the first mentioned, except when another was identified as one of the three favorites in the preferences list at the end of the form or was the first entered for the previous slot), which means that other activities may also have been present (although this was the case for only approximately 3% of entries). In reality, it needs to be acknowledged also that activities were not likely to have been neatly split into whole hour segments, each clearly demarcated. Splitting the day up in this way, however, was thought preferable to making use of narrower divisions such as 15 min used with 12- to 15-year-olds by Curr et al. (1962; 1964), as these smaller time slots may have confounded the youngsters.

Responses for each time slot were coded into one of 44 categories (including 'nothing' where this was actually written and 'missing' where there was no response or the response was unclassifiable). Each of these activities (i.e., responses) was coded also with reference to with whom it was carried out, in terms of 6 categories (viz., 'unclassifiable', 'alone', 'with family', 'with friends', 'with family and friends', and 'with sitter' (this information is not included in the analysis given here). All data were entered into an Excel spreadsheet.

FINDINGS

Patterns of Leisure Behavior

Table 4.2 gives a breakdown of the categories into which children's diary entries were coded. There are 27 discrete headings plus an 'other' category that includes all those activities for which entries for any of the

TABLE 4.2
Frequency of Activities for each Phase

Activity	1994	%	1995	%	1997	%	1998	%
active participation in sports	156	2.03%	207	3.00%	177	3.13%	69	1.61%
activities with animals	203	2.64%	182	2.63%	166	2.94%	76	1.77%
clubs/brigade/guides	143	1.86%	61	0.88%	65	1.15%	24	0.56%
cycling	122	1.59%	138	2.00%	82	1.45%	55	1.28%
drawing/crafts	60	0.78%	85	1.23%	39	0.69%	49	1.14%
dressing/undressing	115	1.49%	97	1.40%	69	1.22%	49	1.14%
eating	1032	13.41%	895	12.96%	668	11.82%	492	11.45%
gardening/outdoor jobs	172	2.23%	163	2.36%	94	1.66%	90	2.09%
go to town/shopping	97	1.26%	114	1.65%	42	0.74%	41	0.95%
homework	74	0.96%	91	1.32%	44	0.78%	83	1.93%
indoor hobbies	1	0.01%	179	2.59%	48	0.85%	13	0.30%
indoor jobs	216	2.81%	28	0.41%	186	3.29%	142	3.30%
listening to radio/music	92	1.20%	99	1.43%	92	1.63%	54	1.26%
other	336	4.37%	245	3.55%	170	3.01%	149	3.47%
party/trips/family outings	116	1.51%	148	2.14%	52	0.92%	73	1.70%
reading books	298	3.87%	250	3.62%	272	4.81%	191	4.44%
sleeping	1391	18.07%	1170	16.94%	880	15.57%	845	19.66%
Sunday school/church/ Kingdom Hall	90	1.17%	127	1.84%	98	1.73%	74	1.72%
swimming	63	0.82%	18	0.26%	84	1.49%	8	0.19%
traveling	226	2.94%	170	2.46%	124	2.19%	96	2.23%
unorganized indoor play	248	3.22%	170	2.46%	100	1.77%	154	3.58%
onorganized outdoor play	450	5.85%	426	6.17%	155	2.74%	184	4.28%
visiting peers	137	1.78%	92	1.33%	55	0.97%	44	1.02%
visiting relatives	146	1.90%	110	1.59%	108	1.91%	81	1.89%
walking	156	2.03%	116	1.68%	56	0.99%	21	0.49%
washing	572	7.43%	580	8.40%	433	7.66%	321	7.47%
watching TV	0	0.00%	155	2.24%	873	15.45%	615	14.31%
watching video	985	12.80%	792	11.46%	419	7.41%	204	4.75%
Totals	7697		6908		5651		4297	

phases did not exceed 1% of the available time slots in any year (viz., 'aimless wandering', 'at work', 'being visited by peers', 'boating', 'dancing', 'fishing', 'hospital visits/illness', 'looking after other children', 'nothing', 'playing computer games', 'playing musical instruments', 'reading comics', 'talking', 'telephone calls', 'visiting graveyard', and 'watching sport'). Each category is listed alphabetically, for each phase of the study.

In interpreting Table 4.2, the reader needs to take into account that there was a total of 30 time slots (all of 1-hr duration except the first on Sunday and the last on all 3 days), over 3 days for each child for each phase. Thus, in the case of the 1994 survey, 269 children were included with 30 entries each giving a total of 8,070. This figure includes 373 'missing' entries where time slots were left blank on the diary forms or where entries were unclassifiable. Percentages indicate the proportion of the time slots that children have allocated to each of the activities (as coded), with the missing entries removed. This has been done so as not to distort the picture in terms of gains and losses that might simply be the product of an increase or decrease in the percentage of missing entries, which were 373 (4.62%) in 1994, 832 (10.75%) in 1995, 529 (8.56%) in 1997, and 713 (14.23%) in 1998. In what follows, the percentage of time slots is given with the missing entries removed, unless an indication is made otherwise.

The second phase survey, carried out in late November and early December 1995 (when only CNN was available) shows that the impact of TV was then relatively minor. Only 2.24% of time slots were spent 'watching TV', although it was already the 13th most often identified activity. However, by the time of the third phase survey in February 1997, when both CNN and M-Net were available, the impact was considerable, with watching TV taking up 15.45% of the available time slots as the second most frequently identified activity with sleeping the first.

Possible displacement effects can be seen by means of a direct comparison between the first pre-TV phase in 1994 and the third phase in 1997, when watching TV reached its highest level. Those activities that lost more than 1% of the available time slots over that period were 'watching video' (5.39%), 'unorganized outdoor play' (3.11%), 'sleeping' (2.5%), 'eating' (1.59%), 'unorganized indoor play' (1.45%), and 'walking' (1.04%). However, there were other 'gainers' additional to watching television, including 'active participation in sports' (1.1%), 'reading books' (0.94%), 'indoor hobbies' (0.84%), 'swimming' (0.67%) and 'Sunday school/church/Kingdom Hall' (0.56%), to name only those with gains of over 0.5%.

However, some of these shifts are no longer in evidence when we take account of the findings from the fourth phase survey. It is noticeable that, comparing February 1997 and October 1998—4 months and 24 months, respectively, after the introduction of multiprogram, two-channel satellite television—we find that 'watching TV' dropped from 15.45% to 14.31% of the available time slots, a fall of 1.14%. Among the other activities that experienced a fall, the biggest was recorded by 'watching video' (from 7.41% to 4.75%), with 'active participation in sports' (3.13% to 1.61%), 'swimming' (1.49% to 0.19%), and 'activities with animals' (2.94% to 1.77%) being the other categories that fell by more than 1%. Those categories that gained by more than 1% were 'sleeping' (15.57% to 19.66%), 'unorganized indoor play' (1.77% to 3.58%) and 'unorganized outdoor play' (2.74% to 4.28%). The 'other' category stayed fairly stable as a proportion of the overall time but it is perhaps worthy of note that it includes 'playing with computer games,' which, although it had increased from 0.5% of the time slots available in 1997 to 0.98% in 1998, had only just overtaken the level of 0.96% it had obtained in 1994.

It is apparent that three of the categories that lost most from 1994 to 1997 (viz., 'unorganized outdoor play', 'sleeping', and 'unorganized indoor play') experienced gains from 1997 to 1998, with only 'unorganized outdoor play' remaining below its 1994 level, whereas the other three continued to experience losses ('watching video' losing 2.67%, 'eating' 0.37%, and 'walking' 0.5%). All of the biggest gainers from 1994 to 1997 experienced losses from 1997 to 1998, viz. 'watching TV' (from 15.45% to 14.31%), 'active participation in sports' (3.13% to 1.61%), 'reading books' (4.81% to 4.44%), 'indoor hobbies' (0.85% to 0.3%) and 'swimming' (1.49% to 0.19%), although 'Sunday school/church/Kingdom Hall' remained at much the same level (falling from 1.73% to 1.72%). It must be acknowledged that some of these falls were very slight (and of little significance given the small numbers involved). However, it is worthy of note that 'active participation in sports' and 'swimming' had by 1998 dropped to below their levels in 1994.

First Preferences

Such changes in the patterns of leisure activities are obviously influenced by children's likes and dislikes. The diary–survey collected information on which activities on each of the days were enjoyed 'most of all', 'next most', and 'third most'. Again, a comparison of the position in 1994 (pre-TV) and 1997 (when 'watching TV' was at its highest level) makes the direction of shifts in leisure-time activity most apparent. Thus, in terms of first preferences, although 'watching TV' had reached the lofty position of 25.19% of such choices by 1997 (excluding the 'missing' entries), gains had also been recorded by 'active participation in sports' (gaining 4.24% of the total), 'reading books' (3.11%), 'swimming' (1.29%) and 'indoor hobbies' (1.11%). The biggest 'losers' were 'watching video' (losing 13.55% of the total), 'unorganized outdoor play' (6.14%), 'eating' (2.2%), 'unorganized indoor play' (2.07%), 'visiting peers' (1.98%), 'clubs/brigade/guides' (1.68%), and 'party/trips/family outings' (1.21%).

However, some interesting shifts also took place between 1997 and 1998 in activities 'enjoyed most of all'. Table 4.3 gives first preference totals for the 3 days combined for each year, 'missing' responses having been removed from the calculations.

Here, the most dramatic falls were experienced by 'active participation in sports', from second in 1997 to eighth in 1998 (11.85% to 4.42% of first choices for the 3 days) and by 'swimming', from joint 6th in 1997 to joint

TABLE 4.3
First Preferences 1997 and 1998 in Rank Order

Third Phase Activity	1997 Number	%	Fourth Phase Activity	1998 Number	%
watching TV	136	25.19%	watching TV	67	18.51%
active participation in sports	64	11.85%	watching videos	29	8.01%
reading books	37	6.85%	unorganized outdoor play	26	7.18%
watching videos	35	6.48%	reading books	23	6.35%
unorganized outdoor play	31	5.74%	cycling	19	5.25%
activities with animals	25	4.63%	unorganized indoor play	18	4.97%
swimming	25	4.63%	sleeping	17	4.70%
unorganized indoor play	22	4.07%	active participation in sports	16	4.42%
cycling	21	3.89%	indoor jobs	15	4.14%
gardening/outdoor jobs	18	3.33%	party/trips/family outings	12	3.31%
clubs/brigades/guides	14	2.59%	activities with animals	11	3.04%
sleeping	11	2.04%	clubs/brigades/guides	10	2.76%
visiting relatives	10	1.85%	drawing/crafts	10	2.76%
eating	9	1.67%	eating	10	2.76%
playing computer games	9	1.67%	playing computer games	10	2.76%
walking	9	1.67%	Sunday School/church/		
listening to radio/music	8	1.48%	Kingdom Hall	10	2.76%
looking after other children	8	1.48%	gardening/outdoor jobs	9	2.49%
indoor jobs	7	1.30%	go to town/shopping	9	2.49%
drawing/crafts	6	1.11%	homework	9	2.49%
indoor hobbies	6	1.11%	listening to radio/music	5	1.38%
party/trips/family outings	5	0.93%	visiting relatives	5	1.38%
nothing	4	0.74%	playing musical instruments	4	1.10%
Sunday School/church/			visiting peers	4	1.10%
Kingdom Hall	4	0.74%	hospital visits/illness	3	0.83%
go to town/shopping	3	0.56%	walking	3	0.83%
visiting peers	3	0.56%	indoor hobbies	2	0.55%
dancing	2	0.37%	telephone calls	2	0.55%
travelling	2	0.37%	nothing	1	0.28%
watching sport	2	0.37%	swimming	1	0.28%
at work	1	0.19%	travelling	1	0.28%
fishing	1	0.19%	watching sport	1	0.28%
playing musical instruments	1	0.19%			
talking	1	0.19%			
Totals	540	100%	Totals	362	100%

28th in 1998 (4.63% to 0.28%). The loss of popularity of these activities is reflected in their similar losses in terms of their share of children's time as recorded in the diary entries. It is interesting to note that over the same period 'reading books' had retained a very similar share of first preferences (third in 1997 with 6.85% and fourth in 1998 with 6.35%). Paradoxically, 'watching video' gained in the number of times it was chosen as the activity enjoyed most of all (fourth in 1997 with 6.48% and second in 1998 with 8.01%), but lost in the time slots where children indicated that they had been undertaking that activity (from 7.41% to 4.75%).

Changes Over Time: The Cohort Studied Over the First Three Phases of the Study

However, measures of the displacement effects of the introduction of television are perhaps best made by looking at the changing patterns of choice exercised by the cohort who went through the first three stages of the process (the 9- to 10-year-olds in Year 2 in 1994, the 10- to 11-year-olds in Year 3 in 1995 and the 11- to 12-year-olds in Year 4 in 1997). These children were not included in the 1998 survey as they had left their middle schools. Table 4.4 gives the leisure pursuits recorded for each of the first three phases by the same 73 children, here identified as the '1994-97' cohort. 'Missing' entries of 100 in 1994 (4.57%), 123 in 1995 (5.62%) and 165 in 1997 (7.53%), are excluded.

By 1997, for these children who had experienced the first three phases, 'watching TV' had reached the lofty position of 23.27% of activities 'enjoyed most of all' (compared to 25.19% for all children at that time). However, for this 1994–1997 cohort, considerable gains in popularity over the three years had also been recorded by 'active participation in sports' (gaining 5.69% of the total of first preferences compared to an increase of 4.24% for all children at that time), 'listening to radio/music' (2.97% compared to 0.01% overall), 'reading books' (2.79% compared to 3.11% overall), 'swimming' (2.79% compared to 1.29% overall), and 'looking after other children' (2.48% compared to 0.68% overall). The biggest 'losers' for them were 'unorganized outdoor play' (losing 12.03% of the total compared to 6.14% overall), 'watching video' (9.99% compared to 13.55% overall), 'gardening/outdoor jobs' (3.61% compared to 0.27% overall), 'unorganized indoor play' (3.15% compared to 2.07% overall), 'visiting peers' (2.6% compared to 1.98% overall) and 'cycling' (2.23% compared to 0.25% overall).

Nineteen of the children from this 1994–1997 cohort indicated that they had watched television for at least 20% of the available time slots over the 3 days in 1997 (up to a maximum of 40%). For these 'heavy viewers',

TABLE 4.4
Frequency of Activities for Each Phase for the 1994–1997 Cohort

Activity	First Phase		Second Phase		Third Phase	
	Number	%	Number	%	Number	%
active participation in sports	45	2.15%	60	2.90%	74	3.65%
activities with animals	40	1.91%	50	2.42%	52	2.57%
clubs/brigade/guides	28	1.34%	21	1.02%	38	1.88%
cycling	49	2.34%	45	2.18%	31	1.53%
drawing/crafts	24	1.15%	32	1.55%	10	0.49%
dressing/undressing	32	1.53%	30	1.45%	26	1.28%
eating	291	13.92%	249	12.05%	238	11.75%
gardening/outdoor jobs	42	2.01%	38	1.84%	14	0.69%
go to town/shopping	21	1.00%	32	1.55%	14	0.69%
homework	19	0.91%	12	0.58%	23	1.14%
indoor hobbies	0	0.00%	68	3.29%	13	0.64%
indoor jobs	49	2.34%	0	0.00%	73	3.60%
listening to radio/music	17	0.81%	37	1.79%	36	1.78%
other	69	3.30%	84	4.06%	57	2.81%
party/trips/family outings	19	0.91%	38	1.84%	19	0.94%
reading books	84	4.02%	79	3.82%	93	4.59%
sleeping	424	20.29%	343	16.59%	325	16.05%
Sunday school/church/ Kingdom Hall	21	1.00%	13	0.63%	29	1.43%
swimming	15	0.72%	2	0.10%	59	2.91%
travelling	78	3.73%	58	2.81%	59	2.91%
unorganized indoor play	56	2.68%	76	3.68%	18	0.89%
onorganized outdoor play	167	7.99%	153	7.40%	51	2.52%
visiting peers	36	1.72%	34	1.64%	22	1.09%
visiting relatives	28	1.34%	34	1.64%	44	2.17%
walking	35	1.67%	23	1.11%	19	0.94%
washing	153	7.32%	193	9.34%	162	8.00%
watching TV	0	0.00%	47	2.27%	260	12.84%
watching videos	248	11.87%	216	10.45%	166	8.20%
Totals	2090		2067		2025	

'watching TV' was indicated in 29.8% of the available time slots over the 3 days (excluding 'missing' responses). Other gainers for this group of 19 children since 1994 (increasing their percentage of the total by more than 0.5%) were 'visiting relatives' (2.37%), 'active participation in sports' (0.85%), 'clubs/brigade/guides' (0.71%), 'watching sport' (0.55%), 'looking after other children' (0.55%), 'Sunday school/church/Kingdom Hall' (0.54%), 'swimming' (0.53%) and 'indoor jobs' (0.51%). For the same group, losers of more than 1% were 'watching video' (9.71%), 'sleeping' (6.14%), 'eating' (4.03%), 'unorganized outdoor play' (3.94%), 'cycling' (1.87%), 'gardening/outdoor jobs' (1.68%), 'activities with animals' (1.51%), 'walking' (1.49%) and 'unorganized indoor play' (1.13%).

The 39 children who took part in the surveys in both February 1997 (as 9- to 10-year-olds) and October 1998 (as 11- to 12-year-olds), after a further 20 months of being exposed to television, are worthy of particular attention. For these children, the biggest fall in the level of activity for any category was 'watching TV' (from 17.95% to 15.02% of their total time slots minus 'missing' entries of 95 in 1997 and 165 in 1998), followed by 'watching video' (from 7% to 4.38%) and 'unorganized indoor play' (from 2.88% to 1.39%). The biggest gainers were 'sleeping' (16.37% to 20.3%) and the catchall 'other' category (from 2.33% to 3.98%). In terms of activity 'enjoyed most of all' (taking out of the reckoning the 'missing' responses), 'watching TV' dropped from 31.4% to 20.5% of favorite activities chosen over the 3 days, but remained the most popular. Interestingly, in 1997 the second favorite activity was 'active participation in sports' (at 9.8%, dropping to 4.6% in 1998), whereas in 1998 this was 'reading books' (9.1%, rising from 2.9% in 1997).

Comparisons of Children of the Same Age

However, for the whole (1994) cohort followed for the first three years, the differences in reported leisure activities might well be explained by the process of maturation, the changing patterns of behavior that occurred as children progressed from Year 2 to Year 4. In order to take account of possible maturational effects, we need to compare different children of the same age at each of the points in the process (although even this is not strictly comparing 'like with like' because these are not the same individuals, and aspects of the culture and environment other than the availability of television may have changed over the period). Thus, comparing the leisure activities of 9- to 10-year-olds in both 1994 (before television was introduced) and 1997 (when a wide range of television programs was available and 'watching TV' was at its height), we find that television had encroached upon a considerable proportion of the time slots (17.22% in 1997). However, other activities had also increased their proportion of the total, viz. 'reading books' (1.07%), 'Sunday school/church/Kingdom Hall' (0.99%), 'listening to radio/music' (0.9%) and 'indoor hobbies' (0.57%), to name those gaining by more than 0.5% of the total. Losers by more than 1% of the total were 'watching video' (4.72%), 'unorganized outdoor play' (4.4%), 'eating' (2.95%), 'sleeping' (2.72%), 'walking' (1.8%) and 'travelling' (1.17%).

A comparison of the leisure activities of 10- to 11-year-olds in 1994 and 1997 shows that, by 1997 when 'watching TV' was at its highest level, television had taken 14.8% of the time slots available. Other activities had also increased their proportion, with 'activities with animals' (1.59%), 'active participation in sports' (1.16%), 'indoor hobbies' (1.02%) and 'gardening/outdoor jobs' (0.84%) gaining by more than 0.5% of the total. However, it is noticeable that the proportion of time slots the 10- to 11-year-olds claimed to watch television (14.8%) was lower than that for 9- to 10-year-olds (17.22%). Furthermore, all the other activities that made gains of more than 0.5% of the total for 10- to 11-year-olds were different from those that did so for 9- to 10-year-olds. The list of activities that lost more than 1% of the total, however, includes most of the same items, viz. 'watching video' (5.44%), 'sleeping' (4.36%), 'unorganized outdoor play' (3.11%), 'eating' (2.34%), 'unorganized indoor play' (1.77%), 'visiting peers' (1.57%), 'travelling' (1.42%), 'visiting relatives' (1.33%), 'clubs/brigade/guides' (1.22%), 'party/trips/family outings' (1.15%) and 'swimming' (1.15%).

For 11- to 12-year-olds in 1997 (the 73 in the '1994–1997' cohort plus five newcomers who were not present in 1994), 13.3% of the time slots were taken by 'watching TV' when this activity reached its highest level in 1997. Again, other activities also increased their proportion of the total by more than 0.5% compared to 1994, viz. 'swimming' (gaining 2.54% for this age group, although being a 'loser' for 10- to 11-year-olds), 'active participation in sports' (1.33%), 'reading books' (1.3%), 'indoor hobbies' and 'indoor jobs' (both 0.64%). Losers by more than 1% of the total of available time slots were 'watching video' (6.39%), 'unorganized indoor play' (2.26%), 'gardening/outdoor jobs' (1.9%), 'unorganized outdoor play' (1.71%), 'sleeping' (4.36%), 'activities with animals' (1.17%) and 'eating' (1.01%). Worthy of note is that the proportion of the time slots taken by 'watching TV' was lower (13.3%) for 11- to 12-year-olds than for both 10- to 11-year-olds (14.8%) and 9- to 10-year-olds (17.22%).

Viewers and Nonviewers

An alternative way to consider the possible displacement impact of television is to make a comparison between the 151 children who claimed to watch at least some television in the third phase (1997) when 'watching TV' had its greatest incidence, and the 55 who did not. Compared to those who watched no television, viewers were less likely to spend time 'watching video' (4.93% less of the total), 'listening to radio/music' (1.98%), 'swimming' (1.89%), 'washing' (1.45), 'dressing/undressing' and 'reading books' (1.3%), to pick out those activities where negative differences were greater than 1%. Overall, for those who watched television, viewing took up 20.8% of the total. Interestingly, the only other activity which was more than 0.5% greater as a proportion of the whole for these children than for those who watched no television, was 'active participation in sports' (0.51%).

However, nonviewers are a very special group and it is perhaps fairer to compare those who watched television in 1997 with all children in 1994, before the advent of television. The children who watched television in 1997 spent an average of 20.8% of their time doing so, but also devoted considerably more of their time than children did in 1994 to 'active participation in sports' (1.24% more of the total time slots), 'indoor hobbies' (0.89% more), and 'reading books' (0.61% more). So although those who watched television in 1997 spent less time 'reading books' than nonviewers, they spent substantially more time doing so than children did overall in 1994. The familiar 'losers' in these terms (of more than 1% of the total) are 'watching video' (6.65%), 'unorganized outdoor play' (3.39%), 'sleeping' (2.75%), 'eating' (2.26%), 'unorganized indoor play' (1.6%), and 'walking' (1.22%).

DISCUSSION

The arrival of television has had a considerable impact on the leisure activities of these 9- to 12-year-old children of St. Helena. However, care must be taken not to exaggerate this impact. By 1997 when the children identified an activity as that 'enjoyed most of all' for each of the three days, only approximately a quarter (25.19%) of these choices were for 'watching TV', and by 1998 this had fallen to 18.51%. Even so, 'watching TV' was by far the most popular of choices (the next best being 'active participation in sports' in 1997 with 11.85% and 'watching video' with 8.01% in 1998). Similarly, although by 1997 when 'watching TV' was at its peak, 14.13% of the available time slots over the three days (16 on the Sunday and 7 each on Tuesday and Thursday) were allocated to 'watching TV' (15.45% of the total excluding 'missing' responses), this still represented little more than four time slots per child (i.e. 4.24 from the 30 available). Time spent 'watching TV' very roughly approximated to an average of 4¼ hours of viewing per child or 1 hour on a weekday (Tuesday or Thursday) and 2¼ hours on a Sunday. However, a weekday average of 1 hour and a weekend average of 2¼ hours, would give a weekly total of just 9½ hours. Even if we combine 'watching video' with 'watching TV', which together took 6.27 time slots from the 30 available or 20.91% of the total, this would give an average weekly total viewing time of 14 hours. This compares to an average of 19 hours a week spent just watching television for children of a similar age in the UK (Central Statistical Office, 1994). By October 1998, the amount of time spent watching TV had fallen slightly from even this modest level.

This overall picture conceals extremes. Some children watched very little, if any, television. As already mentioned, 55 of the 206 children (26.7%) who completed diary forms in 1997 did not allocate any of their time slots to 'watching TV'. Others watched it for most of their leisure time. In 1997, ten children (4.85%) claimed to do so for more than 40% of the available time slots (i.e., more than approximately 12 hours over the three days or two and three quarter hours on a weekday and six and a half hours on a Sunday). The heaviest viewer (a 10- to 11-year-old girl) claimed to view television for 23 of the available time slots, ie 76.67% of her leisure time. Using the overall figures, by 1997 only 'sleeping' was identified more often than 'watching TV' (15.57% compared to 15.45% of time slots once 'missing' entries are removed from the total), with 'eating' (11.82%) the only other activity above 10%.

It is a difficult task to deduce what activities television viewing displaces. The most straightforward answer is probably 'watching video.' St. Helena is an unusual case of video preceding television, whereas most other studies examine the impact of the reverse sequence (for references and a research report see Lin, 1992). Discounting 'missing' entries, 'watching video' was 12.8% of time slots in 1994, but only 7.41% in 1997, a fall of 5.39% of total time indicated. However, 'watching television' grew by more than this amount (15.45%) and, as described earlier, there were several other gainers including 'active participation in sports' (1.1%), 'reading books' (0.94%), 'indoor hobbies' (0.84%), 'swimming' (0.67%) and 'Sunday school/church/Kingdom Hall' (0.56%), to name only those with gains of over 0.5%. Paradoxically, when 'watching video' fell to 4.75% of time slots in 1998 (a fall of 2.66% since 1997), this coincided with a 1.14% fall for 'watching TV' over the same period. No evidence here, then, of a decline in 'watching TV' coinciding with a return to 'watching video' as a former favorite. This would concur with the predictions of Mutz et al. (1993) and Himmelweit et al. (1958), given the functional similarity of the two activities, but conflicts with the findings of Murray and Kippax (1978) that would lead us to expect a return toward pretelevision levels of interest in other media.

Other activities that lost more than 1% of the total of available time slots were 'unorganized outdoor play' (3.11%), 'sleeping' (2.5%), 'eating' (1.59%), 'unorganized indoor play' (1.45%), and 'walking' (1.04%). Elsewhere, small gains and losses were spread across a wide range of activities. Similar findings are evident if we compare those who watched television when that activity reached its peak in 1997 with all children in 1994. There is no easily discernible pattern. The 'gainers' listed previously include two activities that could be carried out at home (whilst waiting for a favorite television program to start), viz. 'reading books' and 'indoor hobbies', but the other three ('active participation in sports', 'swimming', and 'Sunday school/church/Kingdom Hall') involve getting out, at least two of them being highly active in nature. The 'losers' other than 'watching video' again indicate a rough balance of the indoor and outdoor.

Comparing findings from this study with those of prior research, as discussed earlier, we can identify considerable differences. Himmelweit et al. (1958) found that among children in the UK the introduction of television led to the decline of radio listening, comic reading, and cinema attendance. Because there were no cinemas in St. Helena at any time during the survey, the third of these is obviously not a factor. 'Reading comics' in St. Helena does not seem to have been a popular activity at any stage and has stayed almost steady as a proportion of total leisure time between 1994 and 1998. 'Listening to radio/music' (a category that included all types of radio listening and any references to listening to music whether or not on the radio) had increased from 1994 to 1997 (from 1.2% to 1.63% of the total) along with 'watching TV', in apparent contradiction to the 'functional similarity' hypothesis, but fell back to 1.26% in 1998. Again, as for 'watching video', the decline in the amount of time spent 'watching TV' from 1997 to 1998 does not appear to have led to more time being spent on other forms of mass media, but rather, the opposite.

Nevertheless, the functional similarity theory appears to fit the experience of 'watching video' up until 1997 as it took a large share of the impact of the advent of live television, i.e. the rise of TV led to the decline of video because these were likely to fill similar leisure needs. Perhaps its functional similarity to 'watching TV' also explains the fact that they both declined together from February 1997 to October 1998 (in accordance with Himmelweit et al., 1958), but only if we can assume that other sorts of needs were becoming more important.

It is true also that the other activities which lost more than 1% of the total between 1994 and 1997, viz. 'unorganized outdoor play', 'sleeping', 'eating', 'unorganized indoor play', and 'walking', can fit into the category of relatively unorganized 'marginal' activities which both Himmelweit et al. (1958) and Mutz et al. (1993) would predict would suffer reductions. One of the contributions of this study is to give details about the nature of such activities and the pattern of shifts that has emerged within them.

In other respects, the 'functional reorganization' theory proposed by Brown et al. (1974), might be thought to fit better the outcomes of the study of St. Helena than that of 'functional similarity.' Thus, the 'functional similarity' theory might have led us to expect a fall in the amount of time devoted to 'reading books' (assuming the needs met by such an activity are in some ways also met by 'watching TV'). However, to a limited extent, the opposite was the case, which suggests that 'reading books' might take on new purposes in these circumstances. It is even possible that some of the time spent 'reading books' might be promoted by 'watching

TV.' However, in other respects the empirical findings from this study are at some variance with those from Brown et al., with 'active participation in sports' in St. Helena experiencing substantial gains comparing 1994 and 1997 in contrast to losses in the number of 'rule governed outdoor activities' in Arisaig. Again, it is worthy of note that when 'watching TV' declined over the period February 1997 to October 1998, so did 'active participation in sports' (from 3.13% of time slots in 1997 to 1.61% in 1998).

The findings of Murray & Kippax (1978) in Australia are similar to those of this study in one important respect, in that activities other than watching television experienced increases after its introduction. However, the pattern of gains and losses is markedly different between the studies, with some of the activities losing the most in Australia being gainers in St. Helena, the most notable of these being 'active participation in sports', although, as noted above, this activity declined along with 'watching TV' between 1997 and 1998.

Williams & Handford (1986) found that television's 'negative impact was greatest for sports, and the effect was stronger for youths than for adults' (p. 183) and that the substantial effect of television was especially great on 'active participation in sports'. The St. Helena findings show that 'active participation in sports' outside of school time by 9- to 12-year-olds actually increased over the period when television was introduced (by 1.1% of the total) up until 1997. Even a comparison between those who watched and those who did not watch television in 1997 in St. Helena shows that viewers engaged in more 'active participation in sports' (3.26% compared to 2.75% of the total).

In this study there appears to be no systematic loss of out of home compared to in-home activities, which the work of Williams & Handford (1986) would imply (although Brown et al. would lead us to expect the opposite). To test this further, all activities were classified as either 'out-of-home' or 'in-home' (removing the 'missing' entries from the calculations). The balance of gains and losses were then calculated, comparing children's responses in 1994 (before TV) and 1997 (when 'watching TV' was at its highest level). The out-of-home activities lost 6.4% of the total overall, with those in-home gaining that proportion. However, it must be remembered that 'watching TV' itself constituted a major part of the increase for in-home activities (an increase of 15.45% of the total). If the gains for 'watching television' are removed from the calculation, the loss for in-home activities is 9.05% of the total, which compares to the loss of 6.4% for out-of-home activities. This finding would seem to fit with the predictions of Brown et al. rather than Williams and Handford. It is unlikely that this phenomenon is created by the fact that only one activity could be entered on the spreadsheet for each time slot, thus leaving out other in-home activities carried out concurrently with 'watching TV'. Re-examination of the forms showed that only just over 1% of entries gave another activity in the same time slot as 'watching TV' and the majority of these referred to 'eating.' However, previous studies referred to in Roberts et al. (1993, p. 199) have given much higher estimates for shared activities, which suggests that there is an element of underrecording in this survey. Nonetheless, it seems that in the case of St. Helena, out-of-home activities tended not to lose out to watching television to a greater extent than other in-home activities and that there is some evidence to the contrary.

As shown above, the patterns of displacement vary according to vantage point, e.g. looking at the shifts experienced over the first three phases by the same cohort, comparing different children of the same age before and after the advent of multiservice television, examining the contrasts between viewers and nonviewers in 1997, comparing 1997 viewers to all children in 1994, and comparing frequencies and first preferences in 1998 with those in 1997. The influence of age-related factors is clearly apparent, with differences between the '1994–1997' cohort and the total population of children who took part in 1994 and 1997 best explained by such factors. For those 'heavy viewers' who were members of the 1994–1997 cohort, the pattern of displacement effects was not very different to that for the others of similar age, although the magnitude of the shifts was in some instances greater for the 'heavy viewers'.

CONCLUSION

On the basis of the evidence collected so far from this 'natural experiment', the children of St. Helena seem to have taken to watching television enthusiastically, but not (for the great majority) to the extent where their lives have become dominated by it, even in 1997 when interest was at its height. There is no evidence of a sizeable shift from common sense notions of what might be considered developmentally or educationally 'desirable' towards 'undesirable' activities (see also the critique of the displacement hypothesis in Roberts et al., 1993). The children of St. Helena have continued to engage in a wide range of out-of-home and in-home activities, with no great shift from one to the other to provide for their viewing time. Overall, in 1997 when 'watching TV' reached an early peak, 9- to 12-year-olds spent somewhat less time in 'unorganized outdoor play,' 'sleeping,' 'eating,' 'unorganized indoor play' and 'walking' (together falling by 9.68% of the total time slots) than their equivalents in 1994, but they also spent a total of 4.12% more of their time slots engaged in 'active participation in sports' (which increased from 2.03% to 3.13% of total time slots), 'reading books' (which increased from 3.87% to 4.81% of total time slots), 'indoor hobbies', 'swimming,' and 'Sunday school/church/Kingdom Hall. The losers were some activities that involve little prior organization, or which like sleeping and eating, already take up large amounts of time where losing a little makes an insignificant difference. But, by far the greatest loser between 1994 and 1997 was 'watching video' (from 12.8% to 7.41%), which apparently experienced the most direct loss of time to 'watching television.' Arguably, this may have been beneficial because much of the live material was news and other programs of educational value whereas videos watched before television arrived were mostly films and prerecorded soap operas.

In terms of the longer-term effects, as evidenced by the differences between the findings from the surveys in February 1997 and October 1998, it would seem that Himmelweit et al. (1958) were right in their prediction that the 'novelty' effect of television would eventually fade. This is indicated by both a fall in the number of time slots filled by 'watching TV' (by 1.14%) and the decline in the proportion of choices it received as the activity 'enjoyed most of all', from 25.19% to 18.51%. However, it is less clear that the activities that have regained lost ground are those which are sufficiently different (functionally dissimilar) from watching television. Overall, the biggest gainers from February 1997 to October 1998 have been 'sleeping,' 'unorganized indoor play,' and 'unorganized outdoor play,' which were amongst the biggest losers between 1994 and 1997. The position is also complicated by the fact that for the period between February 1997 and October 1998, 'watching video' has been a bigger loser than 'watching TV' and that falls in 'active participation in sports', 'swimming', and 'activities with animals' have also been substantial. All the biggest gainers from 1994 to 1997 have experienced losses from 1997 to 1998. However, the shifts in activity are by no means unidirectional. It is clear that the overall pattern of leisure activities of the children of St. Helena is undergoing a process of change in which television viewing habits are merely one factor.

To talk of restoring other leisure activities to levels obtained before the introduction of television is to fail to take into account that the children of 1998 are not those of 1994, and the world, even as experienced in St. Helena, is also not the same in other respects. Nor was it the case that the advent of television merely subtracted from all other activities which, with the relative decline of the 'novelty' effect, are awaiting restoration. Other activities grew as a proportion of children's time concurrently with the arrival of television, such as 'active participation in sports', and 'swimming', which have now fallen back to below their pretelevision levels. The gainers since February 1997 seem to be some, but not all, of the relatively unorganized 'marginal' activities such as 'sleeping', and unorganized indoor and outdoor play, which seemed to suffer the greatest losses between 1994 and

1997 (to television and other expanding activities). Of these, only 'unorganized outdoor play' had not by 1998 reached the levels previously occupied in the pretelevision era. There is no evidence of a repetition of what Murray & Kippax (1978) found in terms of the shift back to levels of interest in other media and social participation activities but this is partly because in St. Helena, these losses had not always been initially sustained.

Overall, then, the advent of television in St. Helena cannot be considered in isolation from other developments. The apparent clash between the global cultures represented by television and the localized cultures prevailing in an isolated community has not taken place in a vacuum. For the children of St. Helena the experience of change is part of the process of growing up and watching television has been another aspect of that experience. Changes are taking place in the pattern of leisure activities of the children of St. Helena. The arrival of television has clearly been an important influence, but we cannot assume it has been the only contributory factor. Further research in St. Helena is necessary to study the wider social context and the next steps in this process of change.

CHAPTER FIVE

Relationships Between Children's Viewing Patterns and Social Behavior

Barrie Gunter
University of Sheffield

Charlie Panting
Tony Charlton
David Coles
Cheltenham and Gloucester College of Higher Education

The study of children's playground activity revealed few changes in patterns of behavior following the introduction of broadcast television to St. Helena. Those results contrasted with findings from earlier similar research in which children from a community that experienced the introduction of television reception for the first time did exhibit increased verbal and physical aggression in their playground behavior after television transmissions began (Joy, Kimball, & Zabrack, 1986) Even if significant shifts in playground behavior had been observed across observations taken before and after television transmissions to the island began, it would be difficult to attribute such behavioral changes to television with any precision. The reason for this is quite simply that these group-level observational data provided no information about the television viewing habits of the individual children observed. Although it is possible to speculate that certain patterns of play monitored after the introduction of television broadcasts seemed to contain activities, not previously seen, that could be modelled on those of television characters, without further explanation from the children, we could never be certain that such reasoning was correct.

This chapter considers whether any evidence emerged to indicate relationships between specific patterns of television viewing and antisocial behavior among young children on St. Helena. Rather than examining group play behavior, the analysis presented here was built on viewing data and social behavior data obtained for individual children. Questionnaire-based ratings of antisocial behavior patterns for specific children were related to their diary-based measures of television viewing. In addition, the programs listed in the viewing diaries were separately analysed for how much violence they contained. This meant that precise measures could be produced for each participating child of amount of televised violence to which they had been exposed in their unique viewing diet.

CAN ADVERSE TELEVISION EFFECTS BE EXPECTED?

There is a long history of concern about the impact of televised violence on children's behavioral development. Research has been carried out around the world on this subject. Much of this research has been conducted in developed communities with large populations. St. Helena presents a different kind of social environment comprising a remote and tightly knit community with levels of crime and particularly violent crime that are lower than in many other westernised societies. This type of society may already be inoculated against social and moral infection from the alien invader of broadcast television. Equally, a relatively low benchmark of antisocial conduct may more readily reveal shifts in conduct of significance than would be found in a community with much higher rates of crime.

In the United States, from where most of the research has derived, major commissions of enquiry into the causes of social violence were equivocal in their conclusions about the role of television in this context (National Commission on the Causes and Prevention of Violence, 1969; Surgeon General, 1972). Later reviews of the social scientific evidence concluded that most of the research does suggest a link between violence on television and aggressiveness in children and teenagers (Huesmann & Eron, 1986; Pearl, Bouthilet, & Lazar, 1982). Major studies funded by the American television networks during the 1970s produced conflicting conclusions about the effects of television-mediated violence or antisocial conduct on the antisocial and delinquent tendencies of viewers (Belson, 1978; Milavsky, Kessler, Stipp, & Rubens, 1982; Milgram & Shotland, 1973).

The voluminous research literature can be reduced to a small number of scientific methodologies. Laboratory experiments have used contrived and artificial conditions in which to observe the direct impact of exposure to media violence on narrowly defined forms of carefully controlled aggression. Field experiments have attempted to transport laboratory procedures into nonlaboratory environments such as institutional contexts, where media exposure can be controlled and participants' behavior in semicaptive environments can be continuously monitored. Field surveys have moved even further into the natural environment of media consumers and measured relationships between television viewing and social behavior tendencies that occurred under normal, unconstrained conditions.

Survey research suffers from its reliance on self-report data from respondents about their past behaviors. These accounts may lack the accuracy and precision to demonstrate the reality of individuals' behavior patterns. Furthermore, survey data are restricted to showing correlations among variables, such as television viewing and aggressiveness, which do not in themselves demonstrate causality. Cause–effect analyses can be conducted in laboratory experiments, but have been viewed by some critics as lacking sufficient ecological or external validity to be of any use (Barker & Petley, 1997; Cook, Kendzierski, & Thomas, 1983; Cumberbatch & Howitt, 1989; Gauntlett, 1995; Stipp & Milavsky, 1988). Critiques of survey-based evidence have found the scientific evidence problematic for methodological reasons associated again with the strength of the measures used to establish television viewing and personal aggressiveness or because of inappropriate interpretation of statistical relationships between these variables (Freedman, 1984, 1986).

During the 1980s and 1990s, more sophisticated conceptual models emerged about viewers' involvement with media violence that revealed modified thinking about the psychological mechanisms that underpin audiences' reactions to such material (e.g., Berkowitz, 1984; Berkowitz & Heimer, 1989). The viewer was regarded as playing a cognitively more active role in the context of media content interpretation and choice of response to media stimuli. Even so, comprehensive reviews of the scientific evidence have emerged from the USA that have reached the broad conclusion that a cause–effect relationship between violence on television and child or teenage aggressiveness has now been demonstrated (Comstock & Paik, 1991; Wilson et al., 1996). This largely political position on a still debatable social scientific conclusion does not entirely square with arguments that more attention needs to be given to the context in which violence occurs on television and to the attributes of the violence itself (Wilson et al., 1996). Specifying the relative significance of these features when applied individually or in combination to the overall strength of impact of media violence portrayals is a matter on which further research is needed. In addition, the reactions of audience members to television content and choices of what to watch may be further mediated by their individual personalities (see Conway & Rubin, 1991; Krcmar & Greene, 1999).

More sophisticated psychological models of the processing by viewers of television content represent important developments that enhance the conceptual and methodological ability of media researchers to understand the nature of the medium's potential effects in complex media environments. Although these models are helpful, they do not reduce the problems of poor ecological validity experienced by most experimental research on the mass media. What is needed therefore is a research design that has conceptual and methodological measurement precision and obtains data on naturally occurring behavior. The ideal environment to study the impact of television is one where the population has had little or no prior exposure to the medium. This phenomenon is a rarity today. Few reachable communities exist in which television is being introduced for the first time. Early research of this sort in "virginal" television communities took place when the medium was in its infancy and indicated effects upon children's allocation of time to tasks and the displacement of certain activities by television viewing (Himmelweit, Oppenheim, & Vince, 1958; Schramm, Lyle, & Parker, 1961). As television spread, later studies were pushed out to study more remote communities left behind in the television era by quirks of geography. In these communities, leisure activity displacement was observed in the short term (Coldevin & Wilson, 1985; Williams, 1986). However, evidence emerged in at least one case that such displacement effects weakened in the longer term (Coldevin & Wilson, 1985).

Displacement effects are not surprising. Television viewing uses time. This time must be found somewhere. This means that certain other activities, whether they involve hobbies, sports pursuits, work, socializing, or sleeping may be expected to have time subtracted from them to make time available for watching television. More worrying were findings that indicated upward shifts in the prevalence of antisocial conduct among youngster following the onset of television in a community that had previously had no television reception (Williams, 1986). The extent to which children in remote communities react with increased aggression upon experiencing television entertainment for the first time may depend on whether they perceive the Western lifestyle and its role models as attractive or whether they regard such things as a threat to their local heritage (Granzberg, 1985).

What these previous studies lack is any detailed measurement of the nature of individual children's viewing experiences. In the Canadian three-community study reported by Williams and her colleagues, behavior changes were measured before and at some point after the introduction of television using (a) group-level observational data where the only treatment variable was the presence of television and (b) data from individual children on estimated television viewing (self-report) and aggressiveness (peer and teacher ratings; e.g., Williams, 1986). It was previously noted that observations of children's playground behavior have limited value in relation to showing effects of television because they provide no indication of how much television the children being observed generally watched. Although individual-level data were reported on children's aggressiveness and television viewing in this study, the viewing data were limited to children's own estimates of how much television they watched in hours before and after school on weekdays and at weekends (Williams & Boyes, 1986). These viewing data provided no indication of the types of content to which the children were exposed. In the context of demonstrating the effect of television on the development of child aggressiveness, it is important to know not just how much television was viewed but also what kinds of programs were viewed. The key question is whether children are attracted to watch programs that provide good or bad behavioral examples.

What is really needed are data at the level of the individual child about his or her particular viewing habits and about the types of program content that have featured prominently in their viewing diet. The value of knowing the kinds of violence to which children were exposed stems in part from long-standing concern about children copying antisocial behavior depicted on the small screen, especially when it is enacted by attractive characters with whom they identify (Bandura & Walters, 1963; Boyatzis, Matillo, & Nesbitt, 1995).

Previous studies that have explored links between exposure to violence on television and children's aggressiveness in natural environments have tended to restrict their measures of television violence exposure to self-reported viewing of programs, endorsed from lists of program titles or recorded in viewing diaries. Programs are then generally classified by independent judges or the researchers themselves as violent or nonviolent (e.g., Eron, Huesmann, Lefkowitz, & Walder, 1972; Huesmann & Eron, 1986; McIntyre, Teevan, & Hartnagel, 1972; Milavsky, Kessler, Stipp, & Rubens, 1982). However, these violence measures usually comprise very generalised ratings of programs rather than detailed analyses of broadcast output. In this study, children's exposure to television violence was measured by merging content analysis data and viewing diary data for the same programs. Thus, the programs viewed by children could be weighted in terms of content analysis-based measures of violence. More precise measures of violence exposure could therefore be derived in terms of more specific units of analysis such as numbers of violent acts or program running time occupied by violence within programs children reported to have seen.

The Research

Questionnaire-based surveys were carried out in two phases, before and after the introduction of satellite broadcast television, to measure children's social behavior tendencies. Television viewing diaries were used to measure children's viewing experience. Pre-TV data were collected among the child cohort in November 1993, when they were attending nursery classes, 18 months before the first satellite television transmissions. Post-TV data were collected in 1998, some three years after television had been introduced, at which point the child cohort were attending first schools. Viewing data were further elaborated by merging content analysis scores for programs listed in viewing diaries that indicated how much violence each program contained. This enabled viewing data to be further weighted by these violence values to provide measures of children's exposure to televised violence.

Participants

A core group of 47 children provided data about their television viewing and were rated by their teachers on social behavior measures before and after the introduction of broadcast television. There were 23 boys and 24 girls who were first surveyed in 1993 when aged 3.5 years and again in 1998 at age 8. This cohort represented the population of children on the island who had reached the nursery school entry age threshold of 3.5 years. This study focused on this young age group because of their predilection to imitate (see Bandura, 1985).

Social Behavior Measures

Two different behavior rating scales were used in this study that had been especially developed for evaluating the behavior of children aged under 5 years and 5 to 8 years. There is considerable overlap between these two scales however in terms of the types of behaviors they assess. Both operate on the principle of having a cutoff score with children scoring above that threshold being classified as manifesting behaviors that require special educational provision.

Pre-television social behaviour data were collected via the Pre-School Behavior Checklist (PBCL) (McGuire & Richman, 1988). The PBCL was developed for two to five-year-olds and comprises 22 items referring to four areas of behavior in group settings: emotional functioning, conduct, social relations, and concentration. With the exception of four items, scoring is based on a choice between three alternatives; no problem (0), possible problem (1), and definite problem (2), giving a total scale score range of zero to 44. Low scores suggest good adjustment. Children scoring above the cutoff point (12 or more) are deemed to require further investigation.

Examples of PBCL items are: "hardly ever concentrates for more than a few minutes on any table play" (2), "concentration varies, sometimes finds it difficult to concentrate on table play" (1), or "has good concentration, usually stays at table play for 10 minutes or more" (0); "frequently very dif-

ficult to manage or control; problems (e.g., defiant, disobedient, or interrupts during group activities); almost every day" (2), "sometimes defiant,
disobedient, interrupts during group time, or difficult to manage" (1), or
"easy to manage and control" (0).

Test–retest reliability is given as 0.88 and the total test has high reported internal coherence (Cronbach's alpha = 0.83). Validity of the test
has been established by comparing PBCL scores with ratings from independent observers and interviews with teachers. In addition, PBCL scores
have been found to identify prediagnosed clinical samples known to be
experiencing behavioral problems (McGuire & Richman, 1986).

The post-introduction of television social behavior was measured using
the Rutter Behavior Questionnaire (RBQ). This instrument was used by
teachers to rate pupil's school behavior and is appropriate for use with
primary age children. The questionnaire used in this study—Child Scale
B—comprised 26 statements to which the teachers could respond "cer-
tainly applies" (scored 2), "applies somewhat" (scored 1), or "doesn't ap-
ply" (scored 0) in the case of each child's inschool behavior. The statements are associated with the most common behavioral or emotional
problems that occur in school settings and the score range is 0–52. Low
scores of 8 and below indicate satisfactory adjustment.

The questionnaire has been shown to have a high test-retest reliability
(0.89) over a 3-month period and can discriminate between children attending child guidance clinics and peers in the general population
(Rutter, 1967). A cutoff score of nine or more suggests a child is likely to
require further assessment. A neurotic sub-score is found by summing the
responses of four statements ("often worried, worries about many things";
"often appears miserable, unhappy, tearful or distressed"; "tends to be
fearful or afraid of new things or new situations"; "has had tears on arrival
at school or refused to come into the building this year"). An antisocial
sub-score is obtained by summing the response of six statements ("often
destroys own or others' belongings"; "frequently fights with other children"; "is often disobedient"; "often tells lies"; "has stolen things on one
or more occasions"; "bullies other children").

Both social behavior tests have been used worldwide over many years
with repeated research evidence testifying to their robustness as indicators of problem behaviors among the age groups for which they were developed (Charlton et al., 1993; Richman & Graham, 1971; Rutter, Tizard,
& Whitmore, 1970; Rutter, Cox, Tupling, Berger, & Yule, 1975).

Television Viewing Measures

A three-day television viewing diary was used to measure children's television viewing (covering Sunday, 29 March, Tuesday 31 March, and Thursday 2 April, 1998). Viewing data were collected on programs broadcast be-

tween the hours of 6.00am and midnight. Programs broadcast between the school hours of 9.00 a.m. and 3.00 p.m. on weekdays were excluded. The diary was divided into 1-hour time slots and the children were required to write in the names of the programmes they had watched, for how long, on what services and at what particular time of day. The diaries were distributed by their school teachers who advised the children to complete the diary immediately after viewing a program. Because broadcast television reception is not available in all parts of the island, the children were also asked to state if they ever watched television elsewhere, for example, at a friend's house. Although a 3-day viewing period is not very long, the researchers were concerned to avoid over—burdening respondents (teachers and children), especially given their involvement in other aspects of the larger project (i.e., provision of data on leisure activities and school performances).

Viewing behavior was classified according to program genre and the level of violence programs contained. Programs were categorized in 6 principal genres; cartoons, films/dramas, sport, magazines, factual programs, and news/current affairs. The violence classification was derived from a separate content analysis conducted on the programs transmitted during the time periods covered by the viewing diary. The content analysis results are reported in detail elsewhere (Gunter & Charlton, 1999). The analysis was conducted on 136 programs, comprising 34 news and current affairs programs, 9 magazine shows, 15 factual documentaries, 36 sports broadcasts, 32 children's cartoons, and 9 films/dramas. There was also one arts program. Using a standard a priori definition, violence was measured in terms of violent acts and amount of program running time occupied by violence. Both of these measures were used to weight viewing diary data. In addition, two other measures, number of violent acts perpetrated by male actors and by female actors were also included as weighting criteria. Children's viewing profiles were therefore classified in terms of total number of violent acts, total number of male and female violent acts, and total amount of violence in program minutes/seconds to which they were exposed.

Procedure

During survey wave one in 1993, prior to the introduction of television, initial contact was made with the child cohort. PBCL (McGuire & Richman, 1988) measures were obtained from the children via ratings made by their nursery teachers (n = 6). Each child's behavior was rated on each of the 22 PBCL items. Five years later in 1998, and three years after the introduction of satellite broadcast television, an expanded child cohort was contacted again during the final year of their first school. The RBQ (Rutter,

1967) was used by school teachers to rate the behavior of each child. At this time, the teachers also handed out and explained to children the use of the television viewing diaries. These diaries were completed by the children themselves when viewing at home.

Use of Television

Just over half the children (55% or 26 out of 47) watched some television on the three diary-keeping days. This evidence does not mean that those children (45%) who did not record any viewing on those days had not watched on other days. For the purposes of this study, however, comparisons were made between viewers and nonviewers on the three criterion viewing days. Among those children who reportedly watched any television during the period of viewing measurement, 13 were boys and 13 were girls.

Children who watched any television at all on the three survey days viewed for an average of 3 hours 10 minutes across that period. Girls were heavier viewers than boys. Boys watched for an average of 2 hours and 36 minutes and girls watched for an average of 3 hours and 51 minutes. The most popular program genre with children was cartoons, occupying more than half the viewing time (55%) of those who registered any television viewing on the three survey days. Most of the additional viewing by girls compared with boys was devoted to watching films.

Viewing of Television Violence

The amount and nature of violence on St Helena television during the three survey days was analysed elsewhere (Gunter & Charlton, 1999). The content analysis data were merged with children's viewing diary data to produce measures of exposure to televised violence for individual children. Among those respondents who registered any television viewing, the mean number of violent acts watched was 95. This figure was higher

TABLE 5.1
Television Viewing by Program Genre (in Minutes)

	All	Males	Females
Cartoons	103.46	80.00	118.13
Films	45.19	22.50	59.38
Sports	14.62	27.00	6.88
Magazines	12.12	12.00	12.19
Factual	10.38	14.00	8.13
News/current affairs	4.04	0.00	6.56
All television	189.81	155.50	211.27

for girls (107) than for boys (75). More acts of violence seen by these children were perpetrated by male actors (n = 59) than by female actors (n = 22). This pattern was true of boys (72 versus 16) and girls (65 versus 29). This pattern was true of boys (51 versus 12) and girls (63 versus 28).

In terms of program running time, the mean amount of violence seen by these children on the three survey days was 7.28 minutes. This figure was higher for boys (10.23 minutes) than for girls (5.43 minutes). Thus, although girls watched more television overall than did boys and were exposed to more individual acts of violence, boys saw a greater quantity of violence in terms of the duration for which violence was present on screen when they were watching.

Relations Between Programs Viewed and Violence Seen

A series of zero-order Pearson correlations were computed between program genres watched and measures of exposure to televised violence for those children who watched any television. Total amount of television watched was positively correlated with the total number of violent acts seen ($r = .83$, $p < 0.001$), the total number of male violent acts seen ($r = .72$, $p < 0.001$), and the total number of female violent acts seen ($r = .68$, $p < 0.001$). Across child viewers, viewing of cartoons was positively correlated with total number of violent acts seen ($r = .91$, $p < 0.001$), total number of male violent acts seen ($r = .90$, $p < 0.001$), and total number of female violent acts seen ($r = .53$, $p < 0.01$). Viewing of films was positively correlated with the total number of female violent acts seen ($r = .68$, $p < 0.001$). Finally, viewing of magazine programs was positively correlated with total number of violent acts seen ($r = .55$, $p < 0.05$), total number of male violent acts seen ($r = .44$, $p < 0.05$), and total number of female violent acts seen ($r = .39$, $p < 0.05$).

For boys, total television viewing positively correlated with total number of violent acts seen ($r = .81$, $p < 0.001$) and total number of male violent acts seen ($r = .83$, $p < 0.001$). Among boys who watched television, cartoon viewing was positively correlated with total number of violent acts seen ($r = .92$, $p < 0.01$) and total number of male violent acts seen ($r = .95$, $p < 0.01$). Also for boys, magazine viewing was positively correlated with total number of male violent acts seen ($r = .77$, $p < 0.02$).

For girls, total television viewing was positively correlated with total number of violent acts seen ($r = .85$, $p < 0.01$), total number of male violent acts seen ($r = .68$, $p < 0.01$), and total number of female violent acts seen ($r = .70$, $p < 0.01$). Among girl viewers, viewing of cartoons was positively correlated with total number of violent acts seen ($r = .90$, $p < 0.001$), total number of male violent acts seen ($r = .87$, $p < 0.001$), and total amount of violence seen in minutes (r = .81, p < 0.001). Viewing of films was positively correlated with total number of female violent acts seen (r = .77, p < 0.001). Magazine viewing was positively correlated with total number of violent acts seen (r = .51, p < 0.05).

TELEVISION VIEWING AND SOCIAL BEHAVIOR

Social behavior measures were taken before the introduction of broadcast television, using the Pre-School Behavior Checklist (PBCL) and again after the onset of television transmissions using the Rutter Behavior Questionnaire (RBQ). The RBQ also provided separate measures of anti-social conduct and neurotic tendencies.

A series of independent samples t-tests were computed to compare viewers and non-viewers in the wave two survey sample on each of the social behavior measures. The mean scores on these tests are summarised in Table 5.2. There were no significant differences between viewers and nonviewers on any of these measures. Large standard deviations relative to the mean scores were observed here. These are accounted for by the fact that most of the children exhibited low mean scores, but a few children scored greatly above the mean. On the PBCL, 14 children scored zero, 22 scored between 1 and 5, 8 scored 6 to 10, and three scored over 10. On the RBQ, 14 children scored zero, 18 scored 1 to 5, 8 scored 6 to 10, and five scored more than 10.

Similar comparisons were made between boys and girls to reveal no significant difference between genders on their pre-TV social behavior scores on the PBCL (boys = 4.35, girls = 4.08; t = .23. ns), no significant between gender differences on post-TV neuroticism scores on the RBQ

TABLE 5.2
Mean Scores on PBCL and RBQ

Viewing Status	PBCL		RBQ total		RBQ Neurotic		RBQ Antisocial	
All Children	Mean	SD	Mean	SD	Mean	SD	Mean	SD
Viewers	3.73	3.69	4.62	5.00	0.92	1.02	1.35	2.21
Non-viewers	4.81	4.56	5.24	4.73	0.90	1.18	1.33	2.01
Boys								
Viewers	4.10	3.90	6.00	5.50	0.60	0.97	2.20	2.70
Non-viewers	4.54	3.26	6.92	5.11	0.77	0.73	2.15	2.19
Girls								
Viewers	3.50	3.67	3.75	4.63	1.13	1.02	0.81	1.72
Non-viewers	5.25	6.39	2.50	2.33	1.13	1.73	0.00	0.00

Note. Standard deviation.

(boys = 0.70, girls = 1.13; t = 1.38, ns), but a significant between gender difference on post-TV RBQ antisocial behavior scores (boys = 6.52, girls = 3.33; t = 2.37, df = 72, p < 0.02), and on post-TV RBQ antisocial behavior scores (boys = 2.17, girls = 0.54, t = 2.87, df = 45, p < 0.01).

Further analyses were computed to examine the relationships between social behavior scores and television viewing measures among children who watched any television across the three survey days in the post-television wave. In these analyses, all television viewing measures were correlated with each social behavior measure. The two sets of social behavior measures, pre-TV and post-TV, were also intercorrelated. Over all children, the PBCL score was nonsignificantly correlated with the total RBQ score (r = .16), the RBQ Neurotic score (r = −.18), and with the RBQ Anti-social behavior score (r = .29). Among boys and girls examined independently, there were still no significant correlations between pre-TV and post-TV social behavior measures.

Only one television viewing measure was significantly correlated with social behavior measures across all child viewers. Viewing of cartoons was correlated with the post-TV, RBQ total score (r = .41, p < 0.05) and the post-TV, RBQ antisocial behavior score (r = .50, p < 0.05). When pre-TV PBCL scores were partialed out, the correlations between cartoon viewing and total RBQ score (r = .92, p < 0.001) and RBQ Anti-social behavior score (r = .44, p < 0.03) remained significant.

Among boys only, none of the television viewing measures was significantly related to pre-TV or post-TV social behavior measures. Among girls only, cartoon viewing was positively correlated with the post-TV RBQ antisocial behavior score (r = .63, p < 0.01). When partialing out pre-TV antisocial behavior scores from the PBCL among these girls, the correlation between cartoon viewing and the post-TV RBQ measure of antisocial behavior remained significant (r = .60, p < 0.02).

TV Violence Viewing and Social Behavior

Further correlations were computed between measures of exposure to television violence and social behavior measures taken before and after the onset of broadcast television. Again, these analyses were computed over all child viewers and among boys and girls separately.

Across all child viewers, significant correlations emerged between the total number of violent acts seen and total post-TV, RBQ score (r = .39, p < 0.05) and the RBQ Anti-social behavior score (r = .45, p < 0.05). There were also significant correlations between the total number of male violent acts seen and total post-TV RBQ score (r = .50, p < 0.01) and with RBQ Antisocial score (r = .56, p < 0.01). No other significant zero-order correlations emerged between television violence viewing measures and pre-TV and post-TV social behavior scores.

Among boys only, total number of violent acts seen was not correlated with and pre-TV or post-TV social behavior measures. Number of male violent acts seen was correlated with pre-TV social behavior scores only ($r = .65$, $p < 0.04$). Among girls only, the total number of violent acts seen was correlated with post-TV RBQ Antisocial behavior scores ($r = .62$, $p < 0.01$). Again among girls only, the total number of male violent acts seen was correlated with post-TV RBQ total social behavior score ($r = .60$, $p < 0.01$) and post-TV, RBQ Antisocial behavior score ($r = .74$, $p < 0.001$).

Among boys and girls, all correlations between television violence viewing measures and the post-TV RBQ measure of antisocial behavior reduced to levels of nonsignificance on introducing controls for pre-TV PBCL measures of antisocial tendencies.

Given the correlations that occurred earlier between viewing of certain categories of program, amount of televised violence seen and social behavior measures, further partial correlations were computed to explore relationships between amount of television violence seen and social behavior when controlling for the effects of viewing certain program genres and between viewing of selected program genres and social behavior when controlling for the effects of amount of televised violence witnessed by children. These analyses were computed over all children only.

In the first set of analyses, first-order statistical controls were implemented for cartoon, magazine, and film viewing while examining correlations between total number of violent acts seen, male perpetrated violent acts, female-perpetrated violent acts, and social behavior measures. On controlling for the effects of cartoon viewing, just two significant relations emerged between total number of violent acts seen and scores on the neuroticism scale of the RBQ ($r = .40$, $p < 0.05$), the number of male-perpetrated violent acts seen, and neuroticism ($r = .50$, $p < 0.01$).

On controlling for the effects of magazine viewing, four significant relations emerged: total number of violent acts seen and total RBQ score ($r = .45$, $p <).02$); total number of violent acts seen and RBQ antisocial behavior score ($r = .40$, $p < 0.05$); total number of male-perpetrated violent acts seen and total RBQ score ($r = .55$, $p < 0.004$); total number of male-perpetrated violent acts seen and RBQ antisocial score ($r = .53$, $p < 0.006$).

Finally, on controlling for the effects of film viewing, five significant relations emerged: total number of violent acts seen and total RBQ score ($r = .55$, $p < 0.004$); total number of violent acts seen and RBQ antisocial score ($r = .59$, $p < 0.002$); total number of male-perpetrated violent acts seen and total RBQ score ($r = .57$, $p < 0.003$); total number of male-perpetrated violent acts seen and RBQ antisocial score ($r = .61$, $p < 0.001$); and total number of female violent acts seen and total RBQ score ($r = .40$, $p < 0.05$).

In the second set of analyses, first-order controls were implemented for numbers of violent acts seen when examining correlations between viewing of cartoons, magazine programs or films, and social behavior scores. Controlling, first of all, for the effects of total number of violent acts seen on television, five significant correlations emerged between program genre viewing and social behavior scores: cartoons and PBCL scores ($r = .41$, $p < 0.04$), cartoons and RBQ neuroticism scores ($r = -.42$, $r = 0.03$); magazine viewing and RBQ neuroticism scores ($r = -.43$, $p < 0.03$); film viewing and total RBQ scores ($r = -.51$, $p < 0.008$); and film viewing and RBQ antisocial scores ($r = -.50$, $p < 0.01$).

Controlling next for number of male-perpetrated violent acts seen, four significant correlations emerged: cartoon viewing and RBQ neuroticism scores ($r = -.51$, $p < 0.009$); film viewing and total RBQ score ($r -.43$, $p < 0.03$); film viewing and RBQ antisocial scores ($r = -.40$, $p < 0.05$); and magazine viewing and RBQ neuroticism scores ($r = -.42$, $p < 0.03$).

Controlling finally for number of female-perpetrated violent acts seen, 6 significant correlations emerged: cartoon viewing and PBCL scores ($R = .43$, $p < 0.03$); cartoon viewing and total RBQ score ($r = .44$, $p < 0.03$); cartoon viewing and RBQ antisocial scores ($r = .54$, $p < 0.005$); film viewing and total RBQ score ($r = -.50$, $p < 0.01$); film viewing and RBQ antisocial scores ($r = -.44$, $p < 0.03$); and magazine viewing and RBQ neuroticism score ($r = -.42$, $p < 0.04$).

Discussion and Conclusion

This chapter examined relations between the viewing patterns of children on St. Helena and changes in their antisocial behavior across a period during which broadcast television was introduced to their environment for the first time. A longitudinal survey was carried out among young children aged 4 to 8 years before and after television transmission began in 1995. The children's social behavior patterns were measured through teachers' reports and verbal self-reports. Viewing behavior was assessed through viewing diaries in which the children kept a record of their own use of television (see Charlton & O'Bey, 1997; Charlton, Gunter, & Coles, 1998). Data were also collected on the nature of the program output on broadcast television in St. Helena, with particular reference to the depiction of violence (see Gunter, Charlton, & Lovemore, 1998; Gunter & Charlton, 1999; and in chap. 3).

The analysis reported in this chapter focused on children who were age 3 to 4 during the first phase of data collection and age 7 to 8 by phase two. The special feature of this study is that it merged data from television content analysis with children's personal viewing reports that they registered in TV diaries covering the same broadcast output as the content analysis. The principal concern of this analysis was to examine relationships between children's social behavior measures (focusing on measures of antisocial behavioral tendencies) and their reported TV viewing, and to utilize antisocial behavior measures taken before and after the introduction of broadcast television. The TV viewing measures included scores for total amount of television watching and for watching of specific programming genres. The measurement of children's TV viewing was further enhanced by combining with viewing data content analysis measures that quantified the amounts of violence contained in the programs the children reported having watched. This procedure yielded more powerful measures of children's TV violence viewing experiences than had those used in previous longitudinal studies of relations between reported viewing of TV violence and antisocial behavior among children (e.g., Joy, Kimball & Zabrak, 1986; Wiegman, Kuttschreuter, & Baarda, 1992).

Following a series of different statistical tests that explored relationships between TV viewing experiences and measures of antisocial behavior, little evidence emerged that watching television had had any deleterious impact on individual children's propensities to misbehave, at least among those children still in elementary school. These findings corroborated group level observations that failed to demonstrate any consistent patterns of behavioral change in children, with specific reference to their play behavior on the school playground, before and after the introduction of broadcast television on St Helena (see chap. 6). The latter data provide only limited evidence for either the presence or absence of an effect of television on children, however, because they were not based on viewing records for the individual children observed. The current analysis, in contrast, did obtain data at the level of the individual child, but could still find no strong evidence that antisocial behavioral tendencies among 8-year-olds who watched television during the period of analysis were linked to their television viewing experiences.

Overall amount of television viewing also emerged as a relatively insensitive measure of potential relations between TV viewing and social behavior. This finding contrasts with earlier research from the 3-community study in Canada which reported that a self-report measure of children's television viewing (in hours per week) did emerge as a significant predictor of physical aggressiveness (Joy et al., 1986). However, this analysis was conducted only among those children in the two communities that had television reception throughout the study, and not among children from the community to which television was introduced.

It has been argued, however, that children's viewing of specific genres of programming may be a more sensitive measure than one of self-

reported hours of viewing where no content preferences are differenti-
ated. The gender of the child also emerged as a key factor. What is
significant in a social sense is that although reported viewing patterns, as
defined by the genres that dominated the child's viewing, were associated with how much screen violence the children had been exposed
to, genre viewing patterns were not strongly linked to antisocial behavior
tendencies. Thus, boys and girls who often viewed cartoons were also
the ones exposed to the largest number of violent incidents. Viewing of
films and magazine programs was also positively correlated with total
number of violent acts seen. Out of the three genres of cartoons, films,
and magazine shows, only cartoons exhibited any relation with social
behavior measures. The nature of these relations differed between boys
and girls. For boys, heavier viewing of cartoons was associated with their
pre-TV scores on teacher-rated antisocial behavior. Among girls, heavier
cartoon viewing was associated with higher scores on the post-TV measure of antisocial behavior. The latter correlation remained significant
when statistical controls were introduced for pre-TV antisocial behavioral
tendencies.

When measures of exposure to TV violence were further enhanced by
merging content analysis data for programs with children's TV diary data
for the same broadcast output, significant correlations emerged over all
children between the total number of violent acts they had seen and both
their pre-TV and post-TV antisocial behavior scores. Once again, different
patterns emerged for boys and girls. Among boys, the number of violent
acts they had seen was significantly linked only to the pre-TV antisocial behavior scores, and not to their post-TV antisocial behavior scores. Among
girls, this result was reversed. The latter relationship for girls, however,
did not survive statistical controls for their pre-TV antisocial behavior
scores.

Given that both the viewing of specific program genres and degree of
exposure to televised violence were independently related to some social
behavior measures, further analyses were run to examine the relations between each of these viewing measures and social behavior, with the other
viewing variable statistically controlled. In doing this, it emerged that on
controlling for the independent effects of cartoon viewing, the amount of
exposure to violent acts on television was significantly related only to the
post-TV measure of neuroticism and not to any measures of antisocial behavior. Controls for the independent effects of watching films, however,
resulted in significant relations between degree of exposure to televised
violence and measures of antisocial behavioral tendencies after the introduction of broadcast television. Turning this type of analysis around and
implementing exposure to televised violence as the statistical control variables, resulted in significant relations emerging between cartoon viewing

and pre-television measures of social behavior and post-television measures of neuroticism. In the presence of the same control variable, viewing of films was related to post-television antisocial behavior tendencies.

These findings indicated that it was cartoon violence rather than that situated in more serious drama, such as films, that was linked to antisocial behavior tendencies among children. Indeed, on controlling for the effects of exposure to televised violent acts, film viewing was negatively correlated with antisocial behavior tendencies. Omitting the effects of viewing televized violence, however, revealed that children who exhibited stronger antisocial behavior tendencies prior to the introduction of television were the most strongly attracted to cartoon entertainment on television. Cartoon viewing per se was linked to the absence of anxiety rather than to the presence of antisocial tendencies in the broadcast television environment.

Magazine show viewing exhibited significant relationships with amount of televised violence witnessed by children and with measures of neuroticism among children. Because magazine programs were found to contain virtually no violence, however, such relations must be regarded as spurious. In fact, they can probably be explained by the fact that magazine viewing was highly correlated with cartoon viewing ($r = .54$, $p < 0.005$), a phenomenon that was in turn a function of program scheduling. Thus, heavier viewers of cartoons were also heavier viewers of magazine shows.

In summary, the results indicate that so far there is little evidence to suggest that a heavier diet of TV violence is linked to concurrent levels of antisocial behavior for the 8-year-olds of St. Helena. Whereas initial signs of such a relation emerged for girls, these disappeared once their pre-existing dispositions to misbehave had been taken into account. For the boys studied, there were no signs that their current TV viewing habits, especially their propensity to watch programs with violence, were linked to their tendency to misbehave. However, for the girls only, there was an indication that viewing of cartoons was linked to a tendency to misbehave during their early school years even after the effects of their pre-TV antisocial behavior had been controlled. Thus, the viewing of cartoons was more powerfully related to antisocial behavior tendencies than was the amount of violence viewed.

In the presence of controls for total number of violent acts seen, however, cartoon viewing was significantly linked to post-introduction of television measures of neuroticism, but not to measures of antisocial behavior. In this case, cartoon viewing was associated with weaker neurotic tendencies. This suggests that, if cartoons are having an effect on these children, it could be something other than the violence content of these programs that is the vital ingredient here. One possibility could be the humor and activity levels within cartoons, ingredients that can be themselves sufficient to cause increased excitability in audiences (Zillmann, 1991; Zillmann & Bryant, 1991).

These results provide another useful piece of the puzzle of children's developing relations with television that is being compiled for this remote TV-naïve community. The current findings, however, must be tempered with some important caveats.

First, television is still a new medium in St. Helena and children on the island have had just a few years experience of it. In time, as the choice of channels grows and the range of material to which the island's audience is exposed expands, the relations observed to date could change. Presently, the island's children have been observed to be among the best behaved in the world (Charlton, Abrahams & Jones, 1996; Charlton, Bloomfield, & Hannan, 1993). Compared to the British secondary school counterparts, secondary school teachers on St. Helena reported fewer instances of problem behavior in classrooms (Jones, Charlton, & Wilkin, 1995) and pupils were recorded as remaining on-task for longer periods of time than their overseas' peers (Charlton, Lovemore, Essex, & Crowie, 1995). This closeknit community has low tolerance for misbehavior among children and has traditionally experienced a crime rate that is lower than most other westernised societies. Crime, however, has not been nonexistent, nor has accompanying fear of crime among certain sectors of the population, as noted by Schulenburg in chapter 2. In the longer term, it will be important to monitor whether any subtle changes in behavior patterns occur that might be linked to television. This continued monitoring could be especially important given the already noted relation between TV violence viewing for boys and their earlier tendencies to misbehave.

Finally, the data collected in this study derive from fairly small samples of children and of television output. Both the content analysis and viewing data derive from just three days of television output. This small sample may be sufficient to yield indicative results but cannot be assumed to be representative of either the nature of television output across the year or of children's year-round viewing experiences. In partially offsetting this weakness in the current methodology, it is perhaps worth noting that year on year changes to the nature of the broadcast television output, either in terms of the genre composition of the broadcast schedules or incidence of TV violence, have not yet been substantial (Gunter & Charlton, 1999). Television services in St. Helena are dominated by thematic channels that present content that varies only within a fairly narrow range. Nevertheless, future research should and will, subject to resources, aim to improve the sampling frame and expand on the TV output analysed.

CHAPTER SIX

Children's Social Behavior Before and After the Availability of Broadcast Television: Findings From Three Studies in a Naturalistic Setting

Tony Charlton
Ronald Davie
Cheltenham and Gloucester College of Higher Education

Barrie Gunter
University of Sheffield

Cilla Thomas
Education Department, St. Helena

This chapter reports on three research studies, each using separate and independent measures of behavior to monitor young children's behavior across the availability of broadcast television on the island of St. Helena. Outcomes from all three studies are consistent in suggesting that social behavior has not altered significantly since the availability of television.

Study 1 involved cohorts of nursery class children (i.e., 3- to 5-year-olds). As with many other investigations of television viewing effects, this study incorporates teachers' ratings of children's behavior around school. Experienced teachers become both skilled at watching over their children's behavior in social settings, and sensitive to any changes that take place. Hence, their vigilance (as well as the judgments they formulate from this activity) can make important contributions to inquiries studying television's effects on young viewers' behavior.

Study 2 investigated 3- to 8-year-olds' freeplay behavior in two school playgrounds in 1994 (before the availability of broadcast TV), and on two post-TV occasions in 1997, and 1998. A notable mark of this study is its singularity. A literature search uncovered just one naturalistic pre- and post-TV study examining children's freeplay behavior (Williams, 1986). However, whereas Williams' study used proximal observations, the St. Helena study is unique in undertaking analyses of freeplay behavior taken from videorecordings of children filmed both before and after the inception of television.

Study 3 centered on a focus group discussion with 16- to 18-year-old students from the island's Prince Andrew Secondary School. Although this discussion was relatively brief and unstructured, interesting and potentially important explanations were put forward to account for some of the findings from the two other studies; findings which on the whole showed little change in children's behavior following the introduction of broadcast television on the island. The students' discussion threw light on potentially mitigating environmental factors.

LEARNING BY OBSERVING

Learning often takes place through copying, modeling or imitating other people's behavior. Emotions, vocational and sporting skills, prosocial and antisocial behaviors, as well as attitudes and values typify the learning which is accessible through this process. Considerable evidence for this learning process has been gathered. For instance, Bandura and colleagues in the 1960s (e.g., Bandura & Walters, 1963; Liebert & Baron, 1972; Bandura, 1986) carried out a great deal of experimental work in laboratory settings on imitative learning by children. The results of this work led them to postulate and then establish a distinction between their subjects' acquisition of a response capability (as a result of observing a model's behavior, mostly on film) and the actual performance of that response. The distinction emerged from the experimental finding that differential patterns of results were found for these potential and actual responses in relation to the experimental variables which were used. For example, Bandura and collaborators discovered that these two response modes differed depending on whether the model was rewarded or punished for her or his behavior. Clearly, subjects were unlikely to practice a behavior that was expected to lead to punitive consequences. Also the complexity of the model's behavior differentially affected the experimental outcome as between these two aspects of imitative learning. (This distinction—largely overlooked in research on the effects of TV viewing—is in our view an important one, and it figures in the theoretical framework which we advance in the St. Helena study.) Thus, it was shown that the "actual performance of imitative response patterns is not always necessary for the learning of them" (Mussen, Conger, & Kagan, 1969, p. 119). Whereas earlier studies focusing on observational learning used real-life and filmed models, later research has turned its attention increasingly on effects of television programming (Wood, Wong, & Chachere, 1991).

LEARNING AGGRESSION FROM VIEWING
TELEVISION VIOLENCE

Since its availability a half century ago, television's popularity has burgeoned, and today in the United States 99.4% of households possess a set (von Feilitzen, 1999). Consequently, much of youngsters' leisure-time is now preoccupied with viewing. By way of illustration, in the UK 4- to 15-year-olds tend to view for around 18.5 hours weekly (Office for National Statistics, 1998). Almost identical figures were reported by Truglio, Murphy, Oppenheimer, Huston, and Wright (1996) for a large sample of 3- to 7-year-olds in the United States. Moreover, Sprafkin, Gadow, and Abelman (1992) estimated that American 18-years-olds will have given more time to viewing than to any other activity apart from sleeping. Nevertheless, this popularity has been responsible—at least in part—for television attracting frequent criticism on grounds not only that observing antisocial acts encourages imitations of them but, also, that viewing displaces time away from other more essential and productive pursuits (see chap. 4). Murders, rapes, robbery, and bullying are among a plenitude of antisocial acts for which television is often blamed. According to McGilvery (1991), concerns about imitational learning are amplified by estimations that real-life mayhem in newsreels and violent portrayals in films and other programs, for instance, mean that: The average American child will see 32,000 murders, 40,000 attempted murders and 250,000 total acts of violence on television before reaching the age of 18 years (p. 3).

Furthermore, because viewers are especially likely to imitate models with prestige (Bandura & Walters, 1963), an added worry is that:

> Good characters frequently are the perpetrators of aggression on TV. A full 40% of the violent incidents are initiated by characters who have good qualities that make them attractive role models to viewers. Not only are attractive characters often violent, but physical aggression is frequently condoned. More than one third (37%) of violent programs feature 'bad' characters who are never or rarely punished . . . (Wilson et al., 1998, p. 71).

CONTRADICTIONS AND INCONSISTENCIES
IN THE LITERATURE

Claims are often reported in the literature—particularly in the United States—that viewing violence leads to antisocial behavior, fear and emotional desensitization, and that any argument over these effects is at an end (e.g., Smith, Nathanson, & Wilson, 1999). At other times, counterclaims are voiced that credible evidence is lacking to underpin contentions of this kind (Gauntlett, 1995), whilst others have paradoxically found increases in aggression to be linked to nonviolent rather than violent programming (Sawin, 1990). Additionally, allegations have been made that some of the research is suspect on epistemological as well as methodological grounds (Barker & Petley, 1997). Thus, the television-viewing effects' literature appears both equivocal and perplexing.

Unquestionably, much of this confusion emanates from the disparate methodologies used to investigate links between viewing and behavior (e.g., laboratory, field, qualitative, and naturalistic). For example, laboratory studies frequently make robust claims that viewing aggression encourages imitations of it (e.g., Bandura, 1986) and weakens restraints about behaving aggressively (Jo & Berkowitz, 1994; Potter et al., 1996). However, whereas these experimental studies allow causal inferences, generalizations of results to real-life settings remain problematic due to tendencies to measure only immediate or short-term behaviors, as well as exclude natural consequences of misbehavior. In contrast, field studies incorporate more realism (e.g., naturally occurring consequences are usually present) although at the expense of diminished experimental controls. Whereas outcomes lend a measure of support to laboratory findings (e.g., Boyatzis, Matillo, & Nesbitt, 1995), results have to be treated cautiously as subjects are frequently atypical (e.g., youngsters from institutions, and others who are initially more aggressive). Finally, the naturalistic, or quasiexperimental, inquiry though lacking experimental controls, provides scope for effects to be observed in real life. Yet results from such studies are themselves often discrepant. By way of illustration, although Joy, Kimball, and Zabrak (1986) found physical and verbal aggressive behavior among pupils in playgrounds increased significantly two years after the availability of television, others have failed to replicate these findings (e.g., Wiegman, Kuttschreuter, & Baarda, 1992) and some have reported increased rates of prosocial behavior after observing TV aggression (De Koning, Conradie, & Nel, 1990). Moreover, too few opportunities have arisen for pre- as well as post-television data collections in naturalistic settings to allow for either convincing claims or counterclaims for television effects.

Aware of the above limitations and contradictions, Gauntlett (1997) argued that if:

> after sixty years of research effort, direct effects of media upon behavior have not been clearly identified, then we should conclude that they are simply not there to be found. (p. 2)

REFUTING SIMPLISTIC LINKS

Conflicting findings and opinion of the above kind strengthen arguments that a single, elementary explanation is insufficient to explain effects of television programming. Most likely, a plenitude of contextual and personal characteristics interact with television exposure to determine if (and if so, how much) imitational learning takes place and whether or not this learning becomes translated into performances.

The significance of contextual variables has been stressed in recent reviews of the literature (e.g., Wilson et al., 1998), and the debate about the potential harm of viewing violence is seldom conducted without reference to contextual risk factors. Huesmann (1986), for example, talks of the child's intellect, social popularity, the extent of her or his identification with TV characters, belief in the realism of program content, and the amount of fantasizing about aggression. Elsewhere, Singer and Singer (1980) referred to dangers of children being overly exposed to violent and arousing programming "through parental laxity or ignorance" (p. 302), and more recently, Wilson et al. (1996) identified a number of other potential mediators of viewing-behavior links including whether the violence viewed is justified or rewarded, and the degree of harm caused to victims.

Other contextual considerations include the social contexts of an individual's upbringing and the ways she or he typically views television and how these directly mediate the learning and performing of imitational behavior. When viewers learn antisocial behaviors from their viewing they may elect not to (or may be dissuaded from attempting to) perform them; resolutions of this type are based chiefly on the predicted consequences of performing a behavior (Rotter, 1966). Predicted consequences can encourage or discourage the practice of learned behaviors regardless of whether they stem from viewing experiences. Younger children however are less experienced and are less expert in making these predictions; consequently their behavior needs constant checking and correcting by adults around them. However, for many children (young and older) the fragmentation of the "traditional family" means they grow up distanced from the watchful and caring eyes of grandparents, aunts, and uncles. Although this loss is difficult to compensate for, Singer, Singer, and Zuckerman (1990) cautioned that for an increasing number of children the television set is replacing the extended family.

Across research studies, similar amounts of imitational learning may occur among viewers. The extent to which this learning is performed is connected to the kind of contextual factors referred to earlier. It is this performance—not the imitational learning which precedes it—which media effects' studies measure, for the amount of learning acquired is difficult, if not impossible, to gauge. In which case, conflicting results from media effects' inquiries may often arise because they show only the extent to which children translate such learning into overt behavior, rather than measure the amount of imitational learning which takes place. As mentioned earlier, with younger children in particular, the degree to which this learning is practiced is inextricably linked to the degree to which families, communities and others check and correct youngsters' behavior. Thus, contextual influences are often more influential (for better or worse) in shaping children's behavior than is simple exposure to television programming. Clearly, reasoning of this kind does not entirely match assertions that the debate about whether televised violence does any harm is over, as claimed by some, mainly North American, researchers (e.g., Smith et al., 1999).

Laboratory experiments offer empirical support to the differentiation between learned and practiced behavior. As already mentioned, Bandura, in 1965, demonstrated that the acquisition of a particular imitative response pattern does not necessarily require the actual performance of the response. He showed that whether such responses are performed depends largely on external stimuli (e.g. the availability of contingent societal sanctions and rewards). However, from a media effects' standpoint, few occasions have arisen to test this reasoning in a naturalistic (i.e., quasiexperimental) setting. This paucity is predictable if regrettable, for the only way of testing this reasoning in an unaltered natural setting is to locate a broadcast television-naïve community with low rates of antisocial behavior and a social milieu which watches over—and undertakes regular successful checks on—youngsters' behavior; and then to monitor behavior across the availability of television. If the milieu remains influential, then television's availability should have little impact upon behavior. Fortunately, the island of St. Helena offers a rare opportunity to undertake an inquiry of this kind, and the three studies which follow, all consider young children's behavior across the availability of broadcast television.

Study 1

TEACHERS' RATINGS OF NURSERY CLASS CHILDREN'S BEHAVIOR BEFORE AND AFTER BROADCAST TELEVISION

Participants

Participants were two cohorts of similar-aged boys and girls (M age = 3–10) attending nursery provision attached to the first schools. Nursery children were selected because their dynamic quest for learning, the amount of time they spend viewing and their proclivity to imitation suggest they may be the most susceptible to television's influences.

The 1993 cohort (28 boys, 31 girls) includes children in nursery education at the beginning of the 1993–1994 academic year. The 1998 nursery class cohort (24 girls, 24 boys) attended classes at the commencement of the 1998–1999 academic year. Over 68% of the 1998 cohort regularly viewed television in either their own, or friends' homes.

Instrument

McGuire and Richman's (1988) Pre-School Behavior Checklist (PBCL) was used by nursery teachers to rate their children's school behavior. The checklist includes 22 items on emotional functioning, conduct problems, social relations, concentration, and speech and language functioning. Most items allow choices between three behavioral descriptions scored 0 (no problem), 1 (possible problem), or 2 (definite problem). Reliability and validity data are provided by the authors. Test–retest reliability is given as 0.88 ($p < 0.0001$) and internal consistency is measured using Cronbach's alpha (0.83). Validity was established through comparing PBCL scores with ratings from observers and interviews with staff.

Procedure

In November 1993, and 1998, nursery teachers completed the PBCL for children in their class. Completed checklists were marked by researchers. A random sample of these pupils ($N = 14$) was rated again by nursery class assistants in 1998 to provide an inter-rater reliability check. The reliability index was calculated by dividing the number of agreed ratings by the total number of ratings (i.e., the sum of agreed + disagreed ratings). In the absence of a directional hypothesis, data from the two cohorts were subjected to two-tailed t-tests.

RESULTS

The PBCL total mean score for the 1993 cohort was 3.2 (median score = 1.0; modal score = 0). Cronbach's α coefficient was 0.85. The 1998 cohort's PBCL total mean score was higher at 4.9 (although there was no overall difference between the two groups which was not likely to be accounted for by chance fluctuations), the median score higher (2.0), whereas the modal score remained the same. Cronbach's α coefficient was 0.89.

Boys' PBCL total mean score increased non-significantly from 4.2 in Phase 1 (pre-TV) to 6.6 in Phase 2 (post-TV). The difference in means for the boys of 2.4 represents an effect size of 0.42, a medium effect according to Cohen's (1992) classification. A post-hoc power analysis for a two-tailed t-test using an alpha level of 5%, the combined standard deviation of 5.74 and the sample size of 52 gave a power of 0.32 suggesting that there was only a 32% chance of identifying an effect of this size. 18 of the 22 cross-sectional comparisons for boys showed no statistically significant changes (see Table 6.1). However, significant differences were noted on the remaining 4 items; 1998 boys were rated as having lower concentration, $t(50) = 2.96$, $p < .05$ and higher activity levels, $t(50) = 2.24$, $p < .05$, and as being both more fearful, $t(50) = 2.25$, $p < .05$ and more inclined to whine, $t(50) = 2.85$, $p < .05$. Girls' PBCL total mean score increased nonsignificantly from 2.30 to 3.00 in Phase 2. There were no statistically significant shifts in any of the 22 cross-sectional comparisons.

Inter-rater reliability in Phase 2 for nursery teachers' and nursery assistants' ratings was 0.86.

DISCUSSION

Forty-four months after the availability of broadcast television, nursery class teachers' ratings of their children's social behavior suggested a continuance of much of the good behavior noted prior to the availability of television in 1993. Across nearly four years of television, there was no significant difference between cohorts' PBCL total mean scores with boys and girls combined. Similar results were found for girls over the same period. Furthermore, no significant differences were noted between the cohorts in the incidence of antisocial behavior (e.g., "fights," "interferes with others," "teases," "difficult to manage," and "destructive"). However, teachers rated the 1998 cohort boys as having significantly poorer concentration, higher activity levels, being more fearful and more inclined to whine.

There is no obvious explanation for these differences, although Singer and Singer (1986b) concluded that their studies of pre-schoolers supported the idea that viewing high action and fast pacing programs can contribute to children's restlessness and fearfulness. Singer and Singer (1986b) argued also that television:

> fosters dependence on external stimulation as a sustaining resource in periods of delay, empty time, or frustration. Thus, at the very least, we can expect heavy viewing to create an increased restlessness in those children who have not already acquired play skills. (1986, p. 109)

Following this line of reasoning, it is feasible that some young viewers, and boys in particular, in St. Helena have become dependent on a fast-pace and undemanding television diet to the extent that, for some of them, nursery class tasks overtax—and so distract—their concentration thus allowing for increased (off-task) activity levels. However, as already mentioned, these increases may be unconnected to television's availabil-

TABLE 6.1
Item Means for Boys and Girls in 1993 and 1998

	Boys					Girls				
	1993		1998			1993		1998		
	M	SD	M	SD	t	M	SD	M	SD	t
Activity level	.15	.36	.48	.65	2.44*	.10	.40	.30	.47	1.67
Not liked by others (peers)	.00	.00	.12	.44	1.36	.03	.18	.00	.00	1.00
Wets	.07	.38	.00	.00	1.00	.00	.00	.00	.00	.00
Soils	.07	.38	.04	.20	.40	.00	.00	.00	.00	.00
Poor concentration	.44	.64	1.00	.71	2.96**	.27	.52	.52	.67	1.52
Difficult to manage	.33	.55	.32	.56	.09	.07	.25	.17	.49	.95
Demands attention	.26	.66	.44	.58	1.05	.30	.60	.17	.39	.93
Speech unclear	.33	.55	.52	.59	1.18	.27	.45	.26	.54	.04
Reluctant to speak to others	.26	.53	.20	.41	.46	.07	.25	.13	.34	.75
Temper tantrums	.22	.51	.24	.44	.14	.13	.35	.09	.29	.53
Not sociable with peers	.15	.46	.16	.44	.10	.03	.18	.17	.39	1.61
Whines	.19	.40	.64	.70	2.85**	.30	.53	.30	.47	.03
Sensitive	.11	.32	.28	.46	1.53	.23	.57	.04	.21	1.69
Fights	.33	.48	.36	.37	.18	.03	.18	.17	.49	1.31
Aimless wandering	.19	.40	.16	.57	.24	.17	.38	.09	.29	.87
Interferes with others	.37	.49	.44	.58	.46	.07	.25	.13	.34	.75
Miserable	.11	.42	.12	.33	.08	.07	.25	.04	.21	.37
Teasing others	.37	.49	.36	.57	.07	.07	.25	.04	.21	.37
Withdrawn from staff	.00	.00	.20	.50	2.00	.00	.00	.04	.21	1.00
Destructive	.26	.45	.24	.44	.16	.03	.18	.04	.21	.19
Fearful	.00	.00	.20	.41	2.45*	.03	.18	.17	.39	1.61
Habits (e.g., sucking, biting)	.00	.00	.04	.20	1.00	.03	.18	.09	.29	.78
Total score on assessment	4.22	4.99	6.56	6.41	1.46	2.30	3.16	3.00	3.33	.77

Note. Scores have been calculated on the assumption of unequal variances.
*p < .05; **p < .01 for two-tailed test.

ity. By way of illustration, the comparatively low viewing rates noted in 1998 (on average around 65 minutes daily), hardly suggest that children, and boys in particular, have so immersed themselves in the new visual experience that they have become adversely affected in the manner outlined by Singer and Singer. Even so, this study's findings in general provide a useful external validity check on those reported elsewhere within the project (e.g., see Study 2 and Study 3).

Study 2

FIRST SCHOOL CHILDREN'S PLAYGROUND BEHAVIOR ACROSS THE AVAILABILITY OF TELEVISION

Participants

Participants in this study were boys and girls in the playgrounds of two of the larger first schools. Their freeplay behavior was video-taped in November 1994 (pre-TV), and again in 1997 and 1998 (post-TV). Over the three filming occasions the combined roll of the two schools averaged 250 3- to 8-year-olds. Young children were selected for this inquiry because compared to their older peers they are less likely to be aware of being observed (Sluckin, 1981), are less self-conscious in playground settings (Pepler & Craig, 1995) and their proclivity to imitation suggests that they are more prone to be affected by their television viewing.

The Setting

School playgrounds in two of the larger first schools were selected for filming. Playgrounds were chosen as they present an ideal venue "for viewing naturalistic peer interactions and processes" (Pepler & Craig, 1995, p. 548).

Videorecordings

In this study, videorecordings were made of young children's playground behavior in the two schools during morning-, lunchtime- and afternoon-break periods. In 1994 and 1997, filming was undertaken within a 2-week period, and in 1998 it was delayed by an additional 2 weeks owing to inclement weather. During filming, a video camera was sited at a corner in each playground, with an operator close by. Videorecordings started when the operator's presence ceased to attract attention. There were four videorecorder placements, at the corner of each diagonal, in order to enable corners of the playgrounds to be covered by the filming. Placements were changed at the beginning of each filming session. The children entered into, and exited from, the video frame at will. School staff continued with their normal supervisory duties during recording sessions, thus offering protection to children on those occasions when children were exposed to harmful or dangerous behavior from peers, thereby lessening any need for intervention by the camcorder operator. There were 968 minutes of freeplay recorded on three filming occasions. The year 1994 produced filming times of 139 minutes for School 1 and 117 minutes for School 2. In both 1997 and 1998, the filming time on every occasion was 178 minutes.

Videorecordings were preferred to proximal observations because they help control for observer drift over time (i.e., beyond the 4 years of the longitudinal study so far), they allow opportunities for repeated analyses of children's recorded behavior where disagreements arise between coders, and they help limit reactivity to adult observations and simultaneous coding responses. Moreover, of all the sources of information available to assess behavior (e.g., questionnaires, semi-structured interviews, observations) the observation procedure is viewed by many as the "Gold Standard" of assessment (Scott, 1996, p. 105).

Coding Behaviors

Videotaped behaviors were coded by trained coders (5 females; 1 male) selected for their experiences in similar coding exercises. Coders participated in 6 hours of training focusing on the observation, coding and recording of videotaped behavior on the Playground Behavior Observation Schedule (see Appendix 1) used for registering occurrences of 26 behaviors (e.g., games; fantasy play; character imitation, antisocial and prosocial behavior) and the number of children involved on each occurrence (i.e., 1, 2, 3–5, and 6+). The schedule is a revised version of the schedule used in an earlier study (Charlton, Gunter, & Coles, 1998). (1994 and 1997 videorecordings were re-analyzed using the revised schedule. 1998 recordings were examined only with the revised schedule.) The schedule included four categories of antisocial behaviors: Gesture, verbal; Kicking, pushing, hitting; Seizing, damaging property; Non-compliant holding, forcing. In training sessions, preliminary discussion with coders centered on inclusive and exclusive examples of behavior in each of the four categories. Using videotapes, coding practice was undertaken followed by discussion on agreements and disagreements.

A coder reliability index was calculated by computing the number of agreements, then dividing by the number of disagreements plus agreements, and multiplying by 100 to give a percentage agreement (Hall, 1974). Agreements were occasions when coders assigned observed behaviors to the same category (e.g., seizing, damaging property) and grouping (i.e., 1, 2, 3–5, 6+). Reliability was established at two levels. First, during training sessions, paired coders viewed identical 30-second time slots from videotapes, and then separately coded observed behaviors using the Playground Behavior Observation Schedule. Comparisons were then made between the types of behavior and group size noted by the two coders. Disagreements were resolved by replaying the particular video sequence and re-coding. Pairings were changed and further reliability checks undertaken. Training sessions continued until agreement levels reached at least 75%, at which point the observers' reliability was considered satisfactory. Across the 27 checks, reliability ranged between 56% and 96%, with a mean of 81%. Secondly, all actual coding sessions (n = 56) were preceded by reliability checks. Coding continued only when the 75% agreement level was reached. Across the 56 reliability checks, agreement levels ranged between 78% and 97%, with an 86% mean.

Coders were blind to the videorecording dates (i.e., if behavior occurred in pre- or post-TV phases). Working in pairs, they coded behavior from each 1 minute video time slots. Pairings were changed so coders spent approximately the same time working with all others; changes took place after a maximum of three hours of coding. Each time slot was timed using either a digital on-screen timer or a cassette "donk" tape. While observing each video sequence, coders made independent notes of observed behaviors, and then compared them. Agreed behaviors—and groupings—were transferred to the PBOS schedule. To aid observations, the television screen was divided using fine string. Each coding session began with coders scanning all behavior presented on the whole screen (step 1). Next the foreground was scanned up to a point where coders found it impossible to identify the gender of the children (if the gender of the children could not be identified, the behavior was not coded). Then, behaviors presented in the left triangle (step 3), and finally, in the right triangle were observed (step 4) and coded. If one coder observed behaviors which the other did not, the time slot was replayed, and checked. When the playground was crowded, replays of each 1 minute time slot could occur up to seven times.

Coded behaviors were entered on the schedule according to the actual behavior, the sex of the behaver(s) (m = boy, f = girl and x = boy and girl) and the number of children involved. Where two behaviors were presented at the same time both behaviors were recorded. For example, two girls walking arm-in-arm were recorded as follows:

Playground Behavior	1	Pairs	3 to 5	6 or more
traveling (walking, running)		g		
hand holding, arm-in-arm		g		

FIG. 6.1.

Only antisocial behaviors are reported here; data on other behaviors are to be reported elsewhere. Antisocial acts involved 4 of the 26 behavior categories listed in the schedule.

RESULTS

The infrequency of the anti-social behaviors on the videotape recordings precluded the use of the Playground Behavior Observation Schedule's 30-second time segments as units for analysis (on average, antisocial behavior occurred in fewer than 20% of the 1 minute time segments). Therefore, data analyses involved computing the number of antisocial behaviors occurring within 30-minute time units (using data from 60 × 1 minute schedule sheets) for the 1994, 1997 and 1998 videotape recordings.

Mean Number of Antisocial Acts Over Years

Over the three years, the overall mean rate of anti-social behavior per 30 minutes (i.e., with years and genders combined) was 13.03 acts (SD = 8.44). Boys committed more of these acts (8.76 events; SD = 6.20) than girls (3.24 events; SD = 2.67). The most common individual category of antisocial behavior was one boy kicking or pushing another boy (5.24 events per 30 minutes). The incidence of mixed group antisocial behaviors (i.e. involving boys and girls) was small and involved only 34 of a total of 430 anti-social acts.

Mean Number of Antisocial Acts Across Years

The overall mean rate of antisocial behavior declined from 16.4 events per 30 minutes in 1994 to 14.5 in 1997, and to 9.0 events in 1998 (see Table 6.2). Boys' mean rates of antisocial behavior decreased from 9.33 in 1994, to 7.00 in 1998. The 1997 mean rate was 10.08. A one-way ANOVA showed no statistically significant changes in antisocial rates either for rates overall (boys and girls combined), or for boys. A Scheffé test identified no differences between groups at the 5% level either for boys and girls combined, $F(2,30) = 2.50$, $p = 0.10$, or for boys alone, $F(2,30) = 0.78$, $p = 0.46$. The Levene test for homogeneity of variance for both boys and girls combined, and for boys alone, confirm that the assumption of equal variances is acceptable. For girls a one-way ANOVA identified significant differences between groups, $F(2,30) = 4.56$, $p = 0.019$, and a post-hoc Scheffé test found a statistically significant difference between the 1994 and 1998 mean rates; 1998 rates were the lower. Caution should be exercised when interpreting the latter results, because the Levene test for homogeneity of

TABLE 6.2
Mean Rates of Antisocial Behavior Across Years

	YEAR					
	1994		1997		1998	
	Mean	Median	Mean	Median	Mean	Median
Overall rate of antisocial behavior	16.44	16	14.50	16	9	8
Overall rate of male antisocial behavior	9.33	7	10.08	11	7	6
Overall rate of female antisocial behavior	4.78	4	3.67	4	1.67	1
Male antisocial behavior: gesture or verbal	0.33	0	0.92	0	0.17	0
Male antisocial behavior: contact, kicking, pushing, hitting	6.22	5	6.42	7	5.08	3
Male antisocial behavior: seizing, damaging property	0.56	0	0.08	0	0.17	0
Male antisocial behavior: non-compliant holding, forcing	2.22	3	2.67	1	1.58	1
Female antisocial behavior: gesture or verbal	0.33	0	0.50	0	0.17	0
Female antisocial behavior: contact, kicking, pushing, hitting	2.22	2	1.92	1	1.25	1

variance suggests that the equal variance assumption should be rejected. Nevertheless, overall it seems safe to interpret the data as showing that for girls the change in mean rates and the spread of antisocial behavior across the three years does show a significant decline.

Analyses were also undertaken across years, with genders separated and combined, for each of the four categories of anti-social behavior (Gesture, verbal; Kicking, pushing, hitting; Seizing, damaging property; Noncompliant holding, forcing). Although there was a general tendency for scores to decline over the four years (see Figures 6.2, 6.3, 6.4, and 6.5), only one category showed a statistically significant change (i.e. noncompliant holding, forcing). For this category, a one-way ANOVA for girls found a significant difference between groups, $F(2,30) = 5.03$, $p = 0.013$. A post-hoc Scheffé test found a statistically significant difference at the 5% level between the 1994 and 1998 rates; girls in 1998 evidenced fewer physical acts of aggression of this type.

DISCUSSION

There is an important aspect that should not be forgotten when interpreting this study's results. It was not possible to link the behavior of the children at the individual level to their personal viewing patterns. Further-

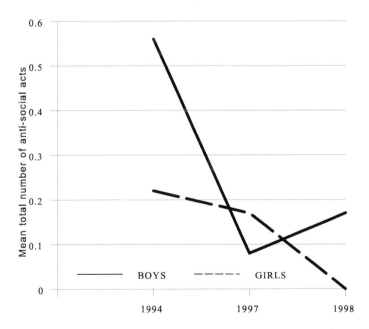

FIG. 6.2. Mean number of antisocial acts per 30 minutes (seizing, damaging property) for girls and boys.

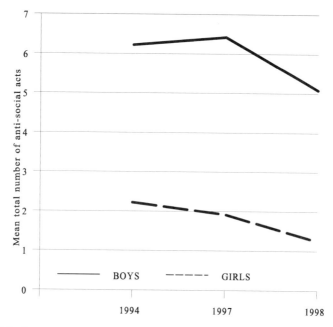

FIG. 6.3.   Mean number of antisocial acts per 30 minutes (kicking, push-ing, hitting) for girls and boys.

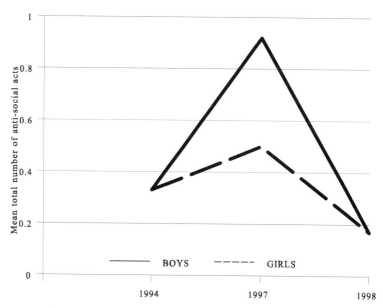

FIG. 6.4.   Mean number of antisocial acts per 30 minutes (noncompliant holding, forcing) for girls and boys.

FIG. 6.5. Mean number of antisocial acts per 30 minutes (verbal abuse) for girls and boys.

more, there could be no control over changes in children's social and cultural environment. Therefore, it is difficult to attribute any observed behavioral changes to television. Regardless, what has emerged from this study is that even at an aggregated group level, few behavior changes were observed that reached statistical significance. Even when changes reached levels of significance these were contrary to the direction of change reported in studies of viewing effects in the large majority of laboratory investigations (e.g. Bandura, 1965; Liebert & Baron, 1973) and in most of the field and longitudinal panel studies (e.g., Boyatzis et al., 1995; Joy et al., 1986). For example, Joy et al. (1986) reported that two years after the inception of television in a small Canadian community, youngsters became significantly more verbally and physically aggressive in 10 of the 12 cross-phase comparisons examined; and for the other two comparisons a similar, though statistically nonsignificant, trend was noted. This was found to be true for both boys and girls, and at more than one age level. Other inquiries have failed to uncover links between television viewing and aggressive behavior (Wiegman et al., 1992) or have found that viewing aggression encourages pro-social behavior (De Koning et al., 1990). Moreover, as long ago as 1984, Freedman's review of research (undertaken when research outcomes and commissioned inquiries, more often than not, showed few doubts on the pernicious effects of viewing violence) led him to conclude that "the available literature does not support the hypothesis that viewing violence on television causes an increase in subsequent aggression in the real world" (p. 244).

Other commentators in the 1980s cast doubt on the consistency and strength of evidence for the effects of television violence based on naturalistic studies of the causes of violence in real life.

In summary, this study observed children's behavior prior to, and following, broadcast television's arrival, using videorecordings of their freeplay behavior. Across that period of time only two significant changes were found and these were in the direction of declining antisocial behavior on the post-TV occasion. Both of these significant changes involved girls' behavior. As such, these findings challenge simplistic notions that viewing television leads to increased levels of antisocial behavior. Study 3 offers one plausible explanation for this largely "no change" situation.

Study 3

STUDENTS' PERCEPTIONS OF TELEVISION'S IMPACT
IN ST. HELENA

Participants

A group of 16- to 18-year-olds (N = 9) scheduled for an A level history seminar at the Prince Andrew Secondary School agreed to give time to discuss television's impact on island life. This group was selected for this study because their age, maturity, and experiences both before and after 18 months of broadcast television, suggested they were well-equipped to reflect on, and make judgments about, any influences of television. The study was used as a pilot inquiry to identify discussion areas ripe for further, and more intensive, inquiry at a later date.

The Setting

The students grouped themselves around classroom desks in their seminar room. They were invited to discuss among themselves their perceptions of broadcast television's impact in St. Helena. No guidance was given to the group. One of the group volunteered to manage the discussion. With the group's approval, their conversation was recorded on audiotape. The exercise began after the teacher and researcher had left the room.

RESULTS

Over an hour's audio-tape recording allowed time for debate on a range of matters covering parental preoccupations with "the box," adult protest marches, a decline in adult socialization at public houses and community centers, imitational behavior on the football pitch, educational benefits of viewing, and wider community effects on youngsters' behavior. Only selected excerpts are included here, in particular those which point to imitational learning (antisocial or otherwise). A fuller account is found in Charlton and O'Bey (1997).

As indicated earlier, the discussion was, at times, directed mostly towards the viewing effects on adults. During the preceding months, two demonstrations had been held on the island. One protest was concerned with teachers' pay, the other opposed to cuts in UK grant-in-aid to the island. To the students these demonstrations were novel, and were a form of group-organized behavior rarely seen in St. Helena. They were in no doubt about where such behavior had been learned:

We've had two protest marches in the last two years.

Two years! Two months you mean.

They got the Governor by the necktie. Big riots and everything. And then you got the teachers marching. But that was a good cause, though. Now where did they learn that?

By watching the News. You always see them doing that sort of thing. It works . . . then you copy it. Like the teachers and the Longwood people (residents of one of the larger inland communities).

There were no marches like that before TV. That's the older generation being influenced by what they see on TV.

Nevertheless, the group acknowledged that this learning was helping islanders to stand up for themselves. For the first time, they could visually access world news instantly, and they could learn ways—and then make use of them—to fend for themselves.

Other comments gave the impression—at least in the early days of viewing—that the TV was left switched on for much of the time. One student found it difficult to undertake her homework reading because "they (the parents?) draw the curtains, and the noise . . . !" Another was wearied by a television diet based chiefly around CNN and world sports. Even so, some admitted that TV (the news, in particular) helped them with their homework. Geography became more alive, and history too. It could supplement their schoolwork by showing "different countries and how different people behave and live."

Who was affected by TV? Students were not short on opinion about who was being affected. It wasn't the younger ones; it was parents and other adults. One student remarked "Look at the old people. You never see them (outside?)." Another added "They don't talk. They're not cooking properly," whereas someone else exclaimed "I think the old people, they never had TV, and now they're getting excited (laughter): but they're getting excited in their own homes, watching TV."

At other times, some admitted that their lives were changing because of the TV. One student confessed that on Sundays she "used to get up and do things" but now she would "get out of bed and switch on the TV." Elsewhere, TV "took over your time. When Newcastle is playing (football) . . . we don't get the homework done." Others agreed with the comment that with the set switched on, "half the time I don't know what I'm watching. I guess I'm just too lazy to turn the TV off."

A preoccupation with the visuality of this new broadcasting medium appeared to be displacing long-running interests in Radio St. Helena. An interest in, and an appreciation of more parochial matters appeared to be on the wane, being replaced by glimpses of famous faces from the pop music and film worlds. Moreover, to some of the group, the attractions (despite repetitions) of frequent news updates were often rated as more popular than Radio St. Helena programs talking about local affairs and interests. More to the point "TV took you to real places" and courtroom dramas such as the O. J. Simpson trial enabled "your eyes to focus on" world events.

When the discussion turned to concerns over violence viewing and its alleged undesirable effects on behavior, the students spoke as one. They all admitted to watching violence on TV. For example:

> When we come home in the afternoon, we watch a bit of violence. . . . Some of the films on Hallmark (one of the TV services) are packed with horror and violence. Blood and brains . . . ugh! And Supersport (another service), that's as bad.

> And the football, that's violence as well. You see players fighting, and kicking . . .

> You watch the wrestling. That's violence for you. And the crowd are screaming them on, telling them to 'Kill them.' And in the cartoons there's always someone bashing someone: that's violence for you. That's murder.

The previous comment was challenged by at least two peers, one of whom raised the point "Yeah, but that's cartoons, isn't it. Not like you get on CNN. You got real murders and real deaths there."

There was some consensus in the group that drama, cartoons and films were of less concern to viewers and that the "worst things to see were wars and riots, and children being crippled by mines and guns." Two or three confessed to not being particularly interested in the news, but ". . . if anything drastic happens . . . then I'm there. Up the front. There was that spacecraft trying to reach the earth. I sat there watching; nothing would shift me."

Even so, students were sensitive to widespread fears that "TV could make you violent." But not in St. Helena, in their opinion. The group acknowledged that this might be the case elsewhere in the world, but as one student commented "I don't think we get more violent or such things . . . Maybe in Britain or America, but not here."

Why was St. Helena the exception, and presumed to be a safe haven from such ill-effects? They acknowledged that it was difficult to indulge in antisocial acts on the island "Because everyone watches you," and "everyone knows you (sigh)." "You've just got to behave when there are people around. And people are around most time on St. Helena." "If you don't behave someone you know will see you. And they'll tell."

DISCUSSION

The students recognized that TV was bringing about changes in the ways in which people, mainly adults, spent their leisure time on St. Helena. For the most part, these shifts were taking place in the ways adults organized their lives; particularly in the ways in which they apportioned their time to the various recreational and other out-of-work activities. At least in the short term, according to the students, adults appeared to be withdrawing into their homes to view supersport and the news, for instance. The "losers" in this displacement exercise often were the socializing activities that took place in the community centers, and public houses. There was a recognition, too, of the educational benefits of viewing. There was evidence, also, in their eyes, of adults learning from their viewing and putting that learning into practice. So the protest marches were seen as a direct manifestation of imitational learning. They were aware, as well, of the potential dangers linked to viewing aggression, but life in St. Helena made it difficult for this to happen on the island. Perhaps, it was the remoteness and isolation of the island and, in particular, the extended family and neighborhood supports that operate within a largely caring community (a composite of many smaller communities) which helped construct a bulwark against untoward effects arising from the viewing of television.

This reasoning is not intended to elevate the island communities as exemplars; for the island has its own peculiar predicaments and concerns (see Foreword to this volume). Many of the parents employed off-island leave their children behind on St. Helena in the care of grandparents and other family members. There are mixed perceptions on the consequences of this arrangement (United Nations Development Programme, 1999). Nevertheless, all societies and communities fall far short of flawlessness, and compared to societies elsewhere St. Helena has managed beneficially to retain links between neighbors, and much of the extended family network (despite some members being absent temporarily). If these characteristics are convincing, then what was the precise makeup of them that seemed to hold more sway over youngsters' behavior than simple exposure to television? Such matters are discussed below.

GENERAL DISCUSSION OF THE THREE STUDIES

Studying the effects of TV viewing can be carried out at the micro level (i.e., gathering data on individuals' viewing habits and personal characteristics in relation to their pre- and post-viewing behavior), or it can be pursued at the macro level, where the data gathered are of an aggregate nature. Both of these approaches have their strengths and limitations. For example, the micro analysis enables a potentially more precise study of individual characteristics but may place limitations on the extent of generalisability from what will be normally a small sample. However, the micro level study also holds out more potential for pinpointing causal factors. On the other hand the use of grouped data, will usually produce a more robust and more reliable analysis in statistical terms but only reveals associations or correlations between variables rather than directly examining possible causation. The two methods are ideally complementary—the one throwing up findings and hypotheses which the other can test. The methodology adopted in this chapter has been of the macro kind and chapter 5 makes use of the micro analysis.

Given the nature of the findings from the three studies and the way in which they conflict with outcomes from many other viewing effects' studies, it is worth noting one other aspect of the St. Helena Project. Gunter, Charlton, and Lovemore (1998), and Gunter and Charlton (1999) undertook a content analysis of violence levels on St. Helena television programs on two weekdays and a Sunday in both 1997 and 1998 (see also chap. 3). They reported that levels of aggressive minutage on the sampling of the island's television programming did not differ greatly from levels on UK television. On the post-TV research occasions in Study 1 and Study 2, the majority of children were regular viewers of programming reported to encourage antisocial behaviors in young viewers including films, news broadcasts and cartoons (Van Evra, 1998; Jenson & Graham, 1995). Therefore, it is unlikely that the absence of significant increases in antisocial behavior noted in Study 1 and Study 2 could be attributed to a vacuity of violent programming of one kind or another.

For the most part, the exemplary social behavior noted in the pre-TV phase was sustained—some 44 months after the inception of television. One possible explanation for this constancy has a potential importance wider than the more focused study of television and behavior. In Study 3, the older students were unconvinced that television would encourage anti-social behavior among youngsters on St. Helena. They said that it was difficult to indulge in anti-social acts on the island "Because everyone watches you . . . everyone knows you. . . . You've just got to behave" (Charlton & O'Bey, 1997, p. 134). Remarks in their discussion, suggested the existence of a "neighborhood watch" on St. Helena, a kind of uncoordinated pastoral network. The vigilance and supervision linked to this network helped shape youngsters' behavior for the better, in homes and schools as well as in the wider community. It helped, also, to check unacceptable behavior. Thinking of this kind is strengthened by Pamela Lawrence's remarks in the Foreword of this book when she reflects on personal expectations on the island. She talks of "mirror images" which families and community reflect back to individuals about their behavior. Consequently, for most islanders there is a obligation to behave as expected. Those who do not, risk the consequences. In like manner, in chapter 2, young students make reference to the informal vigilance whereby it becomes difficult to engage in misbehavior without someone seeing you and informing your parents.

If perceptions of this kind are substantiated by further work of a more empirical nature (which work is now in hand) the possibility emerges that the students' shared opinion suggests one convincing explanation for the continuation of good behavior across the availability of broadcast television. In other words, children's behavior was influenced mainly by supervision (or social control) over their behavior, both with and without television.

A number of social scientists have noted similar 'inducements' at work in small, and mainly rural, communities in the UK and elsewhere. Valentine (1997), for example, makes the point that in communities where most people are known to most others, feelings of being watched over are commonplace because of the "eyes on the street" (p. 144) which keep youngsters and others under surveillance. Likewise, Cohen (1982) spoke of individuality giving way to communality, although Jones' (1999) comment about the reasons for this, namely the "power of gossip" (p. 9) to produce conformity, is unflattering yet with a ring of truth about it. ISCORE (1999) although extolling the merits of small communities, caution that many societies today are in danger of adopting a radical individualization with a concomitant "decrease in civic virtues and care for common good, a decay of family values, [and an] erosion of solidarity between generations."

The significance of family, neighborhood and community factors in shaping behavior has attracted the attention of many others (e.g., Paterson, 1980; Reiss, 1995; Elliott et al., 1996; Laub, & Lauritson, 1999). For example, in two of the disadvantaged communities they studied, Elliot and colleagues (1996) found that high levels of social control, strong informal support networks and a high consensus on community norms and values were influential in the maintenance of good behavior among youngsters. In a similar vein, Laub and Lauritsen (1999) talk of the import of guardianship behavior and "social capital." Social capital they see as being unique, and define it as:

> the extent to which one has others to rely on for assistance and support. In terms of family management, increased social capital means residents share information about children and others in the neighborhood, thereby establishing community norms regarding acceptable and unacceptable behavior. (p. 138)

Whereas neighborhood and community influences are important so too is the family. Paterson (1980), for instance, talks about delinquency and aggression among children being related to their parents' ability to care about, to monitor, to supervise, and to punish the children's antisocial behavior appropriately. More evidence for contextual or situational effects on behavior stem from work by Rutter, Tizzard, and Whitmore (1970) who showed that many children who presented problems in school, did not do so in the home, and vice versa. Given these findings it is understandable why Campbell, Bibel, and Muncer (1985, p. 176) are critical of models of antisocial behavior which 'pay only nominal attention to the social and situational parameters' which govern the social expressions of behavior.

This social control thesis is not new. Situational considerations of this kind form an integral component of social learning (and most other) theories going back, for example, to Mischel (1970) and Rotter (1966). Moreover, the importance of their influence was tested empirically by Potts, Huston, and Wright (1986). They studied effects on young boys' behavior of a systematic manipulation of situational cues (in this case, aggressively and prosocially cued toys) after exposing youngsters to varied television stimuli (i.e., high and low action, high and low violence). Their findings led them to conclude that particular environmental factors, or cues, can expunge the influences of exposure to television violence (although, in terms of the performance of, not the learning of, violent behaviors).

Together, this reasoning and experimental work imply that where environments actively check on and correct children's behavior, television's capacity to adversely affect young viewers' social behavior is lessened or removed, in general. Likewise, television's scope for prompting pernicious effects on young viewers is heightened where parents, neighbors, schools and communities are unwilling, or unable, to provide these checks, and cues. Thus, where parents and others, or the community at large, appear permissive of aggression (either by deliberately ignoring, overlooking, or actively condoning it), they increase chances that children will behave aggressively, with and without television. Regrettably, desirable checks and cues appear now to be less commonplace. In the UK for example, many families—and communities—are now judged to be less able (or inclined) to watch over their children, and it is sometimes even claimed that families, and communities in particular, have become less caring (see Davies, 1997). Without family, neighborhood and community infrastructures to promote and support desirable behavior and values, children are unlikely to be externally reinforced for adopting and practicing those values. Lacking this reinforcement, children can stay baffled about which values to adopt, for often they have access elsewhere to more dubious ones; for example, ones which are readily reinforced within antisocial peer groups; they become more susceptible to what Campbell et al. (1985) refer to as "street norms." Put more simply, deliberations on television's influence on viewers should not overlook the fact that television is embedded in a broad social context, and its influence can be best understood with reference to that context (see Charlton, 1996, 1998). More specifically, homes, schools, neighborhoods, and communities can, in combination, better shape children's behavior than can simple exposure to television, without or with violent programming. This is not to deny that the latter—lacking beneficial societal influences—can have adverse consequences.

Although the three studies' findings are supportive of beliefs that home, school, and community influences can be more persuasive in shaping children's behavior than exposure to television, such thinking does not exclude a recognition of other situational and personal factors that help mediate TV–behavior links. (Clearly, any links between viewing and viewers' behavior are not as uncomplicated as some have argued.) Nonetheless, a mere acceptance of family, school and community influence is of limited value—except perhaps that it may help lessen some of the criticism directed toward the television. However, this value is enhanced with a greater understanding of particular ways in which those influences work. Knowledge and understanding of this kind can inform efforts intended to help maximize, beneficially, children's broad educational experiences. One of these "experiences" could include provision to encourage better parenting skills, whilst others can be linked to promoting awareness of community and citizenship responsibilities. The absence of "good enough parenting" skills (see Charlton, Coles, Panting, & Hannan, 1999), and the loss of a good community ethos, can place insuperable strains on individuals, families and communities. In combination, helping children to become better parents, more responsible citizens and more effective contributors to their community seem to offer more pragmatic ways for limiting pernicious outcomes of television viewing (and other environmental influences), than many of those currently considered in schools and elsewhere.

Thinking of the above kind emphasizes the need for further research for at least four reasons. First, a continuation of the research will help determine whether short-term results persist. Secondly, the continuation of the monitoring of the 1993 nursery class cohort's behavior will help test Centerwall's (1989) theory that media effects become apparent only some 10 years after viewing. Thirdly, more qualitative inquiry into the intuitive "pastoral network" operating within and across families will help identify particular elements of the youngsters' social milieu which encourage the maintenance of good behavior with and without the availability of television. Finally, further inquiry is needed to investigate the lowered concentration levels and raised activity levels found among boys (Study 1) in attempts to determine if they are linked—directly or otherwise—to television viewing.

CONCLUSION

Given that these studies' findings challenge many others' results on television–behavior links it is important to assess particular characteristics and strengths of this study.

First, the research was rooted in an unaltered natural environment, as opposed to an experimental setting. Whereas laboratory and field settings can test the processes through which television affects viewers, the naturalistic study is best able to determine if those effects occur in real life (Comstock, 1983). Results from the playground behavior study, for example, can be generalized to other real-life settings with much more confidence than those emanating from laboratory settings. Second, pre-television measures showed that young children in school were both uncommonly well-behaved (Charlton, Abrahams, & Jones, 1995; Charlton, Bloomfield, & Hannan, 1993) and task-oriented (Charlton, Lovemore, Essex, & Crowie, 1995). It can be argued that there is more scope for good—as opposed to bad—behavior to be affected adversely. In that event, it would be difficult to arrange a more functional baseline against which to test for television effects. Third, the study monitored social behavior across the availability of broadcast television. Most naturalistic (and field) research has been restricted to collecting data after television's arrival, whereas St. Helena offered exceptional opportunities to monitor social behavior before, and after, the inception of television. Fourth, the studies discussed in this chapter used three separate and independent measures of social behavior, the results from which were consistent in suggesting that social behavior had not worsened significantly after the availability of television. The behavior measures used in most other field and naturalistic studies have been restricted to teacher and peer ratings. Observations of playground behavior have been used seldomly (e.g., Singer & Singer; Joy et al., 1986), and a search of the literature failed to uncover any study using videorecordings to monitor playground behavior prior to and following access to broadcast television. The videorecordings utilized in this inquiry allowed scope not only to test for consistency of coding over years, but also to check behaviors where disagreement occurred between coders. Moreover, a degree of objectivity is possible with the observational procedures used in this sub-study which is denied measures dependent on peer and teacher ratings (Williams, 1986). Finally, the island's remoteness and seclusion reduced opportunities for confounding intrusions such as "off-island" guest viewing (i.e., the nearest land mass is several hundred miles away), and possibilities for subject attrition were minimal.

Revised St. Helena Playground Behavior Coding Frame

1 minute of behavior per sheet	Sheet _____			
Playground Behavior	*1*	*Pairs*	*3 to 6*	*6 or more*
Isolated, no involvement				
Observing peers				
Reading, drawing				
Talking				
Traveling (walking, running)				
Skipping, hopping, jumping				
Running around				
Tag				
Ring games				
Rhythm games, clapping, footwork, dancing				
Play equipment, hopscotch, hoops, balls, skip rope				
Football				
Fantasy play, nonaggressive-realism				
Fantasy play, nonaggressive-absurd				
Fantasy play, weapons				
Fantasy play, noncontact, martial arts				
Fantasy play, contact wrestling, holding, capturing				
Antisocial, gesture, verbal				
Antisocial, contact, kicking, pushing, hitting				
Antisocial, seizing, damaging property				
Physical contact, non-compliant holding, forcing				
Prosocial, gesture, verbal				
Prosocial, sharing, turn taking, helping				
Prosocial, consoling, affection				
Hand holding, arm in arm				
Physical contact, rough & tumble, piggy back, lifting				

Note. Record m (male), f (female), or x (mixed) for each behavior within the 1 minute observation. Record all visible behavior every 1 minute. Circle behavior visible for more than 10 seconds. Bracket partially visible behavior.

CHAPTER SEVEN

Where Next?

Barrie Gunter
University of Sheffield

Tony Charlton
Cheltenham and Gloucester College of Higher Education

Daniel Charlton
Institute of Psychiatry, London

This volume reported findings from a research project that set out to determine the impact of broadcast television on a remote island community. The events surrounding the introduction of television broadcasts to St. Helena in 1995 provided a rare opportunity to conduct systematic social scientific research into how people are influenced by television within a naturalistic setting. Conducting research on this topic in a location as difficult to reach as St. Helena presented considerable challenges. As well as problems posed by geography, there were problems of a political nature to be faced. How would a tiny, closeknit community react to be investigated at close quarters by outsiders? As with any social science research about media effects, the project suffered from various limitations. In some instances, these limitations arose because of limited financial resources. In other instances, the research was deliberately constrained by the need to avoid intruding excessively on the lives of the participants being studied.

Another important aspect of this project was the prominence that it attained in print and broadcast media around the world. It attracted high profile media coverage in the United Kingdom, it was featured in newspaper, magazine, radio, and television news reports as far away as the United States, South Africa, Japan, Australia, Greenland, and South America. Some of this coverage was engineered by the project's researchers, and in other instances one medium fed off another in picking up the story. This coverage provided an opportunity to examine the perspective favored by the media in reporting on this type of research. In the event, it was not surprising to find that most of the media's attention focused on the potential harm that may arise from watching television within this remote community. This news angle was to be expected, given the early emphasis that had been placed on the quality of children's behavior on the island as compared to other parts of the world (Charlton, Bloomfield, & Hannan, 1993; Charlton, Jones, & Abrahams, 1995; Charlton, Lovemore, Essex, & Crowie, 1995). However, it would be misleading to say that St. Helena is an island without social problems or crime (see chap. 2). Indeed, UK press coverage was often drawn to the island (independently of this research project) by matters such as a small-scale protest about the British government's proposal to cut its £8 million annual grant to St. Helena (Wigglesworth & Vulliamy, 1997). This chapter considers the significance of this project in terms of its methodological robustness, attraction of media coverage, and opportunities for further investigation into the impact of television.

LIMITATIONS OF THE RESEARCH

Despite the opportunity presented by this "natural experiment," the research was restricted in its scale by a mixture of resource limitations and the need to avoid being too intrusive. Nevertheless, it is important to acknowledge certain weaknesses in this investigation, as is the case with most other research particularly outside the laboratory. There are several elements of this research program for which strengths and weaknesses need to be considered: (a) sampling of children; (b) sampling of television output for content analysis; (c) length of time for which viewing behavior was measured; (d) validity of observation periods; (e) instruments used to measure behavior; (f) controls for prior audiovisual entertainment experience; and (g) research intervention effects.

Sampling of Children

On the first measurement occasion there were around 1,300 children in the island. Different child samples/ populations were taken for the various surveys of time use and social behavior assessment. Furthermore, different age groups were included in the leisure activities and social behavior sections of the study. Time use research was conducted among children aged 9- to 12-years toward the start of this part of the study in 1994. Leisure activity diaries were completed by an initial contact sample of 269 pupils before the introduction of television broadcasts. This sample became gradually eroded during two subsequent, post-TV waves to 258 in 1995, and 206 by 1997. Individual-level social behavior research was conducted mainly among successive populations (i.e., cohorts) of nursery class children. All studies were dependent on the voluntary participation of children, their parents and their teachers. The application of random or quasi-random sampling frames was precluded by the willingness of participants and the need to maintain a non-intrusive scale of operation. Nevertheless, these respondent numbers represented significant proportions (on many occasions, the population) of the total available children in their respective age groups.

Sampling of Television Output for Content Analysis

Small nonrandom samples of television broadcast days were used for the violence content analysis of television output. Three days of broadcasts, covering all major services, were sampled for analysis in two consecutive years. Around 80 hours of programming was analysed on both occasions. It must be acknowledged that these samples are nonrepresentative and may not therefore fully reflect the nature and variety of television output on the island over an entire year. It can be noted, however, that the distribution of programs by genre across the two small television samples showed little variation year-on-year. Because these two program samples were not drawn at exactly the same time of year, this finding does suggest a somewhat stable genre composition to the broadcast television schedules in St. Helena.

Length of Time for Which Viewing Behavior Was Measured

The integration of the children's viewing data with content analysis data on the portrayal of violence for the same sample of programs was a particular strength of this study. Earlier researchers who had conducted similar naturalistic experiments had used very general measures of children's television diets based on self-reported estimates of hours watched per day or nominations of favorite programs from program lists (Himmelweit et al., 1958). A viewing diary method was used in the current study that enabled the researchers to further classify children's television diets in terms of the types of programs they watched. However, even these data cannot reveal the nature of an individual program episode's content. To enhance these viewing measures in ways that would be of particular value when linking viewing to children's social behavior patterns, violence content analysis data were obtained for the television programs covered by the viewing diary. Each child's individual viewing habit could then be weighted for amount of violence. This data integration procedure proved to be worthwhile because the violence exposure measures proved to exhibit the more significant relationships with children's antisocial behavior scores.

The weakness in these data stems from the brevity of the period for which children's television viewing was measured. The diaries used to measure television exposure, covered three-day periods only. Furthermore, all three days occurred within the same week. Such a brief period of viewing measurement poses a number of problems. Three nonrandomly selected days of viewing cannot represent children's viewing across any given year. There is a possibility that children classified for that period as nonviewers may watch as much television as those classified as heavy viewers across the year as a whole. For a variety of reasons some children may have been unable or unwilling to watch television on those three days. However, this could have been due to unusual circumstances. The days on which viewing was measured were a Sunday, Tuesday, and Thursday. There is a possibility, therefore, that nonviewers on those days nevertheless watched television on the other days of that week, while some of those children designated as viewers for those three diary-keeping days, did not watch on any other days of the same week. Further work is needed to establish whether the viewing patterns and viewer/nonviewer distinctions made for children for these 3-day periods do not depart significant from the patterns and distinctions that may be found for other 3-day periods or for longer periods of viewing throughout a given year. Until this validation exercise is completed, a degree of caution must be attached to the current television viewing data. Future research will aim to extend the periods of television output for which program content assessments and viewing assessments are made, and sample more extensively from broadcast output across the year.

Validity of Observation Periods

Children's interactions were observed in school playgrounds during morning, lunchtime, and afternoon breaks. These observations were made over 2-week periods in two of the larger first schools on the island a few months before television reception was introduced and then annually two years after transmissions began. All observations were captured on videotape and analysed in depth at the project's administration centre in the UK. Thus, observations were limited to children aged 3 to 8. This age group was believed to be a particularly interesting one to study given their proclivity to imitation. A coding schedule was developed through which playground behavior was categorized into 26 different types of action or interaction. A number of different types of antisocial behavior were distinguished including verbal hostility, aggressive gesturing, aggressive physical content with another child, and damaging property. There was little evidence of significant changes to patterns of behavior over time, following the introduction of broadcast television.

Whereas one interpretation of such results could be that the presence of broadcast television in these children's home environment was not associated with any observed change in behavior at school, a note of caution is necessary. The data obtained from this observational exercise made largely group-level assessments of children's behavior. Although the behavior of children playing on their own was cataloged in the same way as that of children playing in groups, these observations were then averaged over all the children observed. Moreover, the behavioral data obtained here were not linked back for specific children to their individual television viewing diets. Indeed, no data were collected on how much television or what kinds of programs the observed children normally viewed. In the absence of such data, it is impossible to attribute, with any confidence, changes in observed behavior in the playground to television viewing. Such findings have more general value in confirming broad patterns of children's conduct indicated, by other verbal test measures, where links with individual television diets were made.

Instruments to Measure Social Behavior

The behavioral measures that were linked directly to children's television viewing derived from paper-and-pencil tests in which teachers provided ratings of their pupil's social behavior. These tests comprised checklists of activity descriptions. For each item on the list, the teacher indicated for a particular child, whether this behavior represented a problem he or she would associate with that child. In the television and antisocial behavior study, different tests were used to assess the children at age 3 to 4 years (Pre-School Behavior Checklist) and age 7- to 8-years (The Rutter Behavior Questionnaire). These tests were selected from the literature on the basis of their proven ability to provide effective measures for these age groups (McGuire & Richman, 1986; Rutter, 1967). There was a considerable degree of overlap between these tests in terms of item composition and similarity between the kinds of teacher ratings required.

Both these tests have been demonstrated to have a high test–retest reliability. The issue over which there is less certainty—as with all other occasions where the tests are used—is the way the tests were applied by teachers and the degree of variability of conditions where individual child ratings were made. Furthermore, with adjectival check lists, a question can be raised about whether a teacher's rating of a particular child reflects that child's behavior over a period of time (as the questionnaire requires) or merely reflects the way the child behaves the day when the ratings are made.

Controls for Prior Audiovisual Entertainment Experience

It was known that although broadcast television transmissions were new to the island, the availability of audiovisual entertainment was not. There had been limited prior experience of onscreen entertainment relayed through a television set in those households that had possessed videorecorders (VCRs) before broadcast television was phased in. In the total absence of reliable statistics on VCR ownership and viewing, it was not possible to control for this factor. It is possible that some of the children who participated in this study had prior video viewing experience.

Research Intervention Effects

One point that has been emphasised earlier was the need to ensure that the research project itself was not overly intrusive. The St. Helena community is small and close-knit. This means that whenever anything unusual happens on the island, everyone quickly gets to hear about it. It was essential to avoid contaminating the social environment of the individuals being studied by overburdening them with project-related tasks. Otherwise, this intervention in itself may have been sufficient to invoke certain patterns of responding that were unnatural and not solely the result of the introduction of broadcast television to the island.

MEDIA COVERAGE

It was inevitable that the St. Helena Research Project would attract the fascination of journalists—first, because the media relish discussion of their own importance and influence; secondly, because of the project's far reaching implications for the long-running debate regarding the potential harmful effects associated with portrayals of violence on television.

The project offered a rare opportunity to conduct a naturalistic field experiment on the impact of the world's most prominent mass medium among a television-naïve community. Few such communities still exist. In addition, for many years, important questions have been discussed in the media about the effects of television upon young viewers. This coverage has tended to focus on potential harmful effects associated with portrayals of violence on television.

The level of media interest in the research soon became apparent in 1993 when unsolicited approaches to negotiate exclusive access to the findings were received from BBC *Panorama*, ITV's *World in Action*, two of the leading current affairs programs on British television at the time, and the *Mail on Sunday*. In this context, the question that needed addressing was not whether to involve the media in the St. Helena Research Project, but how to ensure that discussions of the findings in the public sphere were as accurate and well-informed as possible, and conducted in a way that did not reflect badly on the inhabitants of the island of St. Helena.

Subsequently, a schedule was devised to ensure the timely, controlled release of salient information to journalists at regular, designated stages throughout the research process. The aim of this was to inform coverage of the project by proactively placing information in the public domain. It was also a practical way of allocating specific time and resources to manage media interest. Information was promoted in the form of press releases backed up by written question and answer briefings that provided more factual detail on the research process and outcomes. This publicity was initially targeted at UK-based broadsheet media and serious broadcast news outlets, as well as local media on the island of St. Helena itself.

The importance of disseminating factual, written information was well illustrated when, on the basis of one interview, a national newspaper journalist misunderstood the use of video-recordings in the school playgrounds and reported that the project team were using "secret spy cameras" in the playground to monitor children's behavior. This inaccuracy caused understandable concern on the island and provided an early lesson in the need for careful media management. The incident also raises an interesting question about media accountability and the limited available opportunities to redress in inaccurate coverage. Arguably, changes in regulation which obliged newspapers to correct inaccuracies on the same page of the newspaper in which they originally appeared would provide a much-needed incentive for well-balanced, accurate reporting.

The launch of the project was officially announced to the media in 1995. At this point, the information provided concentrated on a description of the situation on St. Helena and the project's aim to monitor the impact of broadcast television on the island's children. At the outset, the key factors mentioned were that the children were naïve to broadcast television, that they were known to be among the most well behaved youngsters in the world. Whereas criminal behavior is not unknown on the island, misdemeanours, delinquency, and other antisocial behavior are relatively uncommon. This paucity is often attributed, in part, to the support individuals receive from their families and the wider community. Furthermore, the lack of electronic entertainment meant that the Islanders had to make their own recreation in the form of games, sports and dances which served to cement a high degree of social cohesion.

The project leader, Tony Charlton, was extensively quoted placing emphasis on the project's objective to consider both potentially positive and negative effects of television. Within the selected quotes was an implicit message that perhaps St. Helena might serve as a beacon to the rest of the world where television consumption was concerned. Given the island's reputation for family-centeredness and good behavior, perhaps the research would uncover a recipe for the control of television's alleged ills. A society that expects and maintains high moral standards, takes child-rearing seriously, and displays low tolerance for antisocial behavior—a value that is regarded as the responsibility of and is endorsed by everyone—may severely restrict the chances that television will lead children astray. In this, there may be an important lesson for the rest of the world. It was therefore with a degree of optimism that the project was greeted at its outset.

An update on progress with the project was issued the following year linked to a conference at which some early findings were unveiled; an illustration of the use of "news pegs" (i.e., a specific event or occurrences) to encourage media coverage. Further regional and national press coverage followed in the June to October period headlining with the project's initial indications that television was, on balance, having more positive than negative effects on the island's children. This finding was highlighted with headlines such as "TV may not be to blame for violence" (*Western Daily Mail*, June 21, 1996), and "St. Helena study shows benefits of television" (*The Times*, 3 August, 1996).

Despite the project's aims to study effects of television upon children's leisure activities as well as their social behavior, the main point of interest for media were the social behavior findings. The first results to be revealed derived from quantitative research involving videorecorded observations of children's behavior at school and involved measures of aggressive behaviors in the playground. The principal feature of this news angle was that it counterpoised the St. Helena results with the more common view that television violence causes harm. In this sense, the St. Helena results were seen as welcome, but also conflicting of the established wisdom. The explanation for this apparent contradiction in results was that St. Helena is a closeknit community where people look out for each other.

A fresh wave of publicity was triggered by the publication of the proceedings of the previous year's conference at which some of the earliest results from the project had been unveiled (Charlton & David, 1997). This coverage appeared predominantly in the local and regional press, but significant national newspaper spreads were received in *The Independent* (21 July, 1997), *The Express* (22 July 1997), and *The Sunday Times* (3 August, 1997). The project even featured in the leader column of *The Express*. Throughout, this coverage continued to report on the favorable and socially positive response children on St. Helena had apparently shown to television. In fact, these reports covered the same ground as the previous year's press coverage. The essence of these findings was again neatly captured in the headlines: "Island airs good news on TV for children" (*Independent*); "Where innocence lives on" (*The Express*) and "Kids turn on the TV for good behaviour" (*Sunday Times*).

In its opinion column, *The Express* referred to the island's children as "remote but controlled." The point that set apart this commentary piece from the other reporting was its strong note of caution over the results. It noted that the observation that the children were no worse behaved after two years of experience of television than before it arrived was based on interim findings. It would be more significant to monitor what would happen in the longer term. The island's community exhibited closer kinship ties than most and this social cohesion was the main reason for television's lack of impact.

The article in *The Independent* was the only one to interview anyone on the island. Susan O'Bey, head teacher at the secondary school, commented on the popularity of the sports channel among children, especially for watching English football, and also noted that television news had superseded radio news. For once, the reporting went beyond the concern with whether watching television made children more badly behaved. To understand the full extent of television's impact in any community, it is essential to establish how the medium is used and what kinds of programs and program ingredients viewers pay most attention to.

The issues of program ratings and the V-Chip were further elements implicitly linked to the project in some news coverage. Such issues had entered the British government's agenda for broadcasting, following the enactment of new legislation in the United States. This legislation mandated that all new television sets from the beginning of the 21st century should have functional V-chips. It required, also, broadcasters to establish a voluntary program ratings system (see Kunkel et al., 1998). The St. Helena findings entered the equation as an illustration of what could be achieved in the way of control over television and its effects via family and community influences.

Further results were released in April 1998 that concerned the impact of television on children's leisure activities and use of time. These were reported in *The Times* (27 April 1998), *The Guardian* (29 April, 1998), and *The Express* (29 April, 1998). Although repeating again earlier observations that no increase in behavioral problems had been noted following the onset of broadcast television, the leisure activities results held more good news. *The Times*, headlining with "Children read more after arrival of TV," reported that the study had found small increases in time spent reading and doing homework two years into the television era as compared with pre-television benchmarks (Frean, 1998). Additional evidence that the children may have grown more gregarious, with television serving as a source of conversation, was also reported.

In contrast, regional press coverage during the same period almost to-
tally ignored the leisure activity findings, despite their significance to any
overall understanding of television's impact upon St. Helena's children.
The *Gloucestershire Echo* (30 April, 1998) headlined with "Boffins say TV
doesn't turn on violence," *The Post* (6 May, 1998) with "Researchers show
island children unaffected by exposure to TV violence"; and the *Western
Daily Press* (6 May, 1998) with "TV cleared over child violence in island
study."

Press coverage continued during 1999 with attention again being
drawn to the implications of the project's findings for the debate about
the effects of televised violence. The quality of this coverage varied. National coverage in *The Express* represented the view that the influences of
television cannot be considered in a vacuum. The impact of television on
young viewers is dependent on their family background and social environment. Whereas all youngsters can be considered potentially vulnerable
to various television effects, some children must be regarded as especially
vulnerable if they come from backgrounds which neglect to check, and supervise, their behavior.

One misrepresentation of this view occurred in a regional newspaper
(*Gloucestershire Echo*, 3 June, 1999) where Tony Charlton, the project's
leader was quoted as saying that parents themselves should shoulder the
blame for children's antisocial behaviors. In a subsequent rebuttal of this
point and clarification of his position, Charlton (1999) presented the view
that the implications of the research findings from St. Helena were that
healthy family, school, and community influences can combine to shape
children's behavior and may, together, prove to be a much more potent
source of influence than is television. Where these resources are lacking, a
situation could be created where television's influences can become more
potent. This must not be seen as the fault of parents who may try hard to
bring up their children to the best of their abilities in a social environment
that offers very little support.

The coverage of the St. Helena project generally conformed to the typical predominant news frame offered by the press when reporting television violence effects, namely conceiving of television violence as a public
health problem. Often, this hazard is depicted as one for which broadcasters must accept primary responsibility (Gunter & Ward, 1997; Hoffner,
1996). Rather than encouraging viewers to switch off (in the same way
campaigns intended to warn people about the dangers of smoking, excessive alcohol consumption, or recreational drug-taking may encourage
those users to give up these habits) journalism's treatment of the television violence problem places the blame firmly at the door of producers
and regulators. In the St. Helena project, however, the allocation of responsibility centered more on the role and importance of parents rather

than of the broadcasters. The subtext of this news frame was that the potential ills of television have not gone away; they have merely been suppressed by, or have been controlled by, other social forces.

The absence of adverse effects of television among children in St. Helena was explained, in part, by the tight-knit, family-oriented community that, generally, provided a warm, supportive, and disciplined developmental environment. Without this environment to depend on, however, children could still fall prey to the evil influences of television. The question that remains unanswered is whether this remote community can withstand the cultural onslaught of television and maintain its high social standards. The issue most of the media coverage has so far not considered is whether the continuance of television will create further appetites and demands, and give further reason for social unrest within a community whose remoteness helps shelters it for the most part from temptations of the "outside world."

WHERE NEXT?

This volume explored the impact of broadcast television upon children in a small, remote community following the introduction of satellite television transmissions in 1995. The findings examined so far have covered the first few years of broadcast television experience on St. Helena. Although no major shifts in children's behavior have been detected so far, this was perhaps not surprising given the unusual qualities of social cohesion, family-centeredness and friendship on this island. In the longer term, however, certain effects of television may gradually begin to be felt. Initially, these may not be manifest in terms of deterioration in social behavior or the abandoning of social and leisure activities that play such an important part in knitting together this close social fabric of the island. Rather, television influences may be much more subtle.

There is little doubt that children on the island have already been exposed to fresh experiences, courtesy of television broadcasts. The medium provides an important source of world news. It brings into children's homes images of far-flung places they may never get to visit themselves. Television will introduce the children of St. Helena to other cultures and lifestyles. Eventually, they will make comparisons between their own situation at home and the alternatives that appear through television to be offered by other societies. Television will also bring new role models for the children. Already the sports channel has proved one of the most popular to watch. Heroes of football, athletics, and wrestling may provide models children may wish to emulate in different ways. In due course, the channel package will no doubt offer popular music channels, allowing children and teenagers to follow the exploits and talents of their favorite music celebrities. Advertising messages will also open up an expanded world of consumerism and possibly desire on the part of the young people of St. Helena.

There is therefore considerable potential, in the longer-term, for television to contribute to changes to St Helenian society. Many of these changes may be beneficial, but some may not. The changes (or absence of them) observed during the very earliest years of the television era on the island may not reflect what will happen in the years to come. In considering what should come next then in terms of further research, it would seem to be important to broaden the base of the study so far to include some analysis of television effects upon children's awareness, knowledge and perceptions of the wider world, on their consumer socialisation, and following up on wider social trends in relation to crime and antisocial behavior, leisure activities, and the continuing significance of the traditional family. It will of course be important to continue monitoring social and behavioral trends on the island because certain behavioral effects of broadcast television may only become apparent in the long term.

To date, little is known about the nature and influence of the environmental conditions that moderate any links between viewing and behavior. Opportunities to investigate these environmental conditions increases the St. Helena project's potential for helping to account for conflicting research findings and apparent contradictions in the research literature. Hence, arrangements are in hand to build on indicative findings from the project by observing and analysing contextual factors within the home and community which may mediate the nature and extent of any effects of TV-viewing on children's behavior. Research of this kind has the potential to contribute a great deal to our knowledge of ways that not only the television, but also other electronic media and interactional media, may affect the user.

The research so far has focused primarily on behavioral changes among very young children. However, it is equally important to consider the impact of the medium on older children who experience a relaxation of parental control as they grow older and seek independence. Because television will bring the world to the island's children in a way never before experienced by them, it will be important to examine whether this has any influences on their personal aspirations. Given their nature, such effects may not become manifest in young people until they reach adolescence and begin to think seriously about the lives they want to live as adults. St. Helena lacks an economy of its own, being largely dependent on British government grants for its income. Will the younger generation of "saints" want more than the island can offer? Will they call for greater opportunities to make the island self-sufficient through economic development programs, or will they simply wish to leave the island and make a life for themselves elsewhere?

The media environment has gone through a period of unprecedented evolution in the past decade. Future research must embrace these developments as they begin to reach and impact on the island. What will be the effects of further growth in television channels, multi-set households, increased penetration of home computers and access to the Internet? How will a close knit social community respond to an influx of entertainment and information sources to which younger generations will undoubtedly demand access? The current multiservice television provision with only two channels can be monitored by parents and children's consumption of these services can be controlled. This situation will change as the number and range of home entertainment sources grows and involves the use of interactive technology with which children will have far greater confidence and familiarity than their parents.

Although the developments identified above represent important topics for investigation, researchers are aware that they must avoid the temptation to attempt too much detailed study of the island. For one thing, the existence of such a research project and the presence of researchers can be strongly felt in such a small community unless handled with sensitivity and skill. Its mere presence may therefore have a significant social and psychological impact in itself. All this means that as much data as possible should be collected using passive, observational, or document research techniques that can be conducted as unobtrusively as possible and that do not place an excessive data provision workload on the island's population.

References

Abelman, R. (1991). Parental communication style and its influence on exceptional children's television viewing. *Roeper Review, 14,* 23–27.

Anderson, B. (1991). *Imagined communities: Reflections on the origin and spread of nationalism* (Rev. ed.). London: Verso.

Baker, C., Davies, N., & Stallard, T. (1985). Prevalence of behavior problems in primary school children in North Wales. *British Journal of Special Education, 12,* 1, 19–26.

Bandura, A. (1965). Influence of models' reinforcement contingencies on the acquisition of imitative responses. *Journal of Personality and Social Psychology, 1,* 589–595.

Bandura, A. (1986). *Social foundations of thought and action: A social cognitive theory.* Englewood Cliffs, NJ: Prentice-Hall.

Bandura, A. (1994). Social cognitive theory of mass communication. In J. Bryant & D. Zillmann (Eds.), *Media effects* (pp. 61–90). Hillsdale, NJ: Lawrence Erlbaum Associates.

Bandura, A., & Walters, R. H. (1963). *Social learning and personality development.* New York: Holt, Rinehart & Winston.

Barker, M., & Petley, J. (1997). *Ill effects: The media/violence debate.* London: Routledge.

Barnes, G. M. (1995, September). Parents can help prevent teen alcohol, drug use: Support, monitoring, key to prevention, study finds. *Research Institute on Addictions: Research in Brief.*

Barnes, G. M., & Farrell, M. P. (1992). Parental support and control as predictors of adolescent drinking, delinquency, and related problem behaviors. *Journal of Marriage and the Family, 54,* 763–776.

Baron, R. A. (1977). *Human aggression.* New York: Plenum.

Belson, W. A. (1978). *Television violence and the adolescent boy.* London: Saxon House.

Berkowitz, L. (1984). Some effects of thoughts on anti- and prosocial influences of media events: A cognitive-neoassociation analysis. *Psychological Bulletin, 95,* 410–417.

Berkowitz, L., & Heimer, K. (1989). On the construction of the anger experience: Aversive events and negative priming in the formation of feelings. In L. Berkowitz (Ed.), *Advances in experimental psychology* (Vol. 22, pp. 1–27). New York: Academic Press.

Borg, M. J. (1997). The structure of social monitoring and the process of social control. *Deviant Behavior, 18,* 273–293.

Bouwman, H., & Stappers, J. (1984). The Dutch violence profile: A replication of Gerbner's message system analysis. In G. Melischek, K. E. Rosengren, & J. Stappers (Eds.), *Cultural indicators: An international symposium.* Vienna: Austrian Academy of Sciences.

Boyatzis, C. J., Matillo, G. M., & Nesbit, K. M. (1995). Effects of the Mighty Morphin Power Rangers on children's aggression with peers. *Child Study Journal, 25,* 45–55.

Bracey, H. E. (1959). *English rural life: Village activities, organisations and institutions.* London: Routledge & Kegan Paul.

Brewer, J. D., Lockhart, B., & Rogers, P. (1998). Informal social controls and crime management in Belfast. *British Journal of Sociology, 49,* 570–585.

Brown, J. R., Cramond, J. K., & Wilde, R. J. (1974). Displacement effects of television and the child's functional orientation to media. In J. G. Blumler & E. Katz (Eds.), *The uses of communications: Current perspectives on gratifications research.* Beverly Hills, CA: Sage.

Buckingham, D. (1993). *Children talking television: The making of television literacy.* London: Falmer Press.

Bushman, B. J. (1995). Moderating role of trait aggressiveness in the effects of violent media on aggression. *Journal of Personality and Social Psychology, 69,* 950–960.

Campbell, A., Bibel, D., & Muncer, S. (1985). Predicting our own aggression: Person, subculture or situation? *British Journal of Social Psychology, 24,* 169–180.

Cannan, E. (1992). *The churches of the South Atlantic Islands, 1502–1991.* Oswestry, UK: Anthony Nelson.

Caron, A. H., & Couture, M. (1977). Images of different worlds: An analysis of English- and French-language television. *Report to the Royal Commission on Violence in the Communications Industry: Vol. 3. Violence in Television, Films and News* (pp. 220–463). Toronto, Canada: The Royal Commission.

Centerwall, B. (1989). Exposure to television as a cause of violence. In G. Comstock (Ed.), *Public communication and behavior* (Vol. 2, pp. 1–59). New York: Academic Press.

Central Statistical Office. (1994). *Social focus on children.* London: Her Majesty's Stationery Office.

Charlton, T. (1996). Television viewing. In K. David & T. Charlton (Eds.), *Pastoral care matters: In primary and middle schools.* London: Routledge.

Charlton, T. (1997). The inception of broadcast television: A naturalistic study of television's effects in St. Helena, South Atlantic. In T. Charlton & K. David (Eds.), *Elusive links: Television, video games and children's behaviour.* Cheltenham, UK: Park Published Papers.

Charlton, T. (1998a). *Exposure to television as a cause of violence? Interim findings from a naturalistic study in St. Helena, South Atlantic.* Cheltenham, England: Cheltenham and Gloucester College of Higher Education, Faculty of Arts and Education.

Charlton, T. (1998b). Reproaching television for violence in society: Passing the buck—Interim findings from a naturalistic study in St. Helena, South Atlantic. *Journal of Clinical Forensic Medicine, 5,* 169–171.

Charlton, T., Abrahams, M., & Jones, K. (1995). Prevalence rates of emotional and behavioural disorder among nursery class children in St. Helena, South Atlantic: An epidemiological study. *Journal of Social Behavior and Personality, 10,* 273–280.

Charlton, T., Bloomfield, A., & Hannan, A. (1993). Emotional and behavioural problems in the middle school population on St. Helena: Prevalence and types of problem behaviour. *International Journal of Special Education, 8,* 101–110.

Charlton, T., & Coles, D. (1997, August). *Teachers' ratings of nursery class children's behaviour before and after the availability of satellite television in St. Helena, South Atlantic.* Paper presented at the International Association of Special Education Conference, Cape Town, South Africa.

Charlton, T., Coles, D., Panting, C., & Hannan, A. (1999). Nursery children's behaviour across the availability of broadcast television: A quasi-experimental study of two cohorts in a remote community. *Journal of Social Behavior and Personality, 14*, 1–10.

Charlton, T., Gunter, B., & Coles, D. (1998). Broadcast television as a cause of aggression? Recent findings from a naturalistic study. *British Journal of Emotional and Behaviour Difficulties, 3*, 5–13.

Charlton, T., Gunter, B., & Hernandez-Nieto, R. (1999). *Children's social behaviour before and after the availability of broadcast television: Analyses of video-recordings of young children's free play behaviour in St. Helena, South Atlantic.* Manuscript submitted for publication.

Charlton, T., Hernandez-Nieto, R., Thomas, C., & Gunter, B. (1999, May). *Young pupils' playground behaviour across the availability of broadcast television.* Paper presented at the 49th Annual Conference of the International Communication Association, San Francisco, CA.

Charlton, T., Lovemore, T., Essex, C., & Crowie, B. (1995). Naturalistic rates of teacher approval and disapproval and on-task levels of first and middle school pupils in St. Helena. *Journal of Social Behavior and Personality, 10*, 1021–1030.

Charlton, T., & O'Bey, S. (1997). Links between television and behaviour: Students' perceptions of TV's impact in St. Helena, South Atlantic. *British Journal of Learning Support, 12*, 130–134.

Cloke, P. J. (1985). Editorial: Whither rural studies. *Journal of Rural Studies, 1*, 1–9.

Cohen, A. P. (Ed.). (1982a). *Belonging: Identity and social organisation in British rural cultures.* Manchester, England: Manchester University Press.

Cohen, A. P. (1982b). Belonging: The experience of culture. In A. P. Cohen (Ed.), *Belonging: Identity and social organisation in British rural cultures.* Manchester, England: Manchester University Press.

Cohen, A. P. (1985). *The symbolic construction of community.* London: Tavistock Publications.

Cohen, A. P. (Ed.). (1986a). *Symbolising boundaries: Identity and diversity in British cultures.* Manchester, England: Manchester University Press.

Cohen, A. P. (1986b). Of symbols and boundaries, or, does Ernie's greatcoat hold the key. In A. P. Cohen (Ed.), *Symbolising boundaries: Identity and diversity in British cultures.* Manchester, England: Manchester University Press.

Cohen, R. (1983). Education for dependence: Aspirations, expectations and identity on the island of St. Helena. In P. Cook (Ed.), *Small island economics.* Manchester, England: Manchester University Press.

Coldevin, G., & Wilson, T. C. (1985). Effects of a decade of satellite television in the Canadian Arctic: Euro-Canadian and Inuit adolescents compared. *Journal of Cross-Cultural Psychology, 16*, 3, 329–354.

Comstock, G. (1983). Media influences on aggression. In A. Goldstein (Ed.), *Prevention and control of aggression: Principles, practices, and research.* NY: Pergamon.

Comstock, G., & Paik, H. (1991). *Television and the American child.* Academic Press.

Conway, J. C., & Rubin, A. M. (1991). Psychological predictors of television viewing motivation. *Communication Research, 18*, 4, 443–463.

Cook, T. D., Kendzierski, D. A., & Thomas, S. V. (1983). The implicit assumptions of television research: An analysis of the 1982 NIMH report on television and behaviour. *Public Opinion Quarterly, 47*, 161–201.

Cross, T. (1980). *St. Helena, including Ascension Island and Tristan da Cunha.* London: David & Charles.

Crow, G., & Allan, G. (1994). *Community life: An introduction to local social relations.* London: Harvester Wheatsheaf.

Cullingford, C. (1992). *Children and society: Children's attitudes to politics and power.* London: Cassell.

Cumberbatch, G., & Howitt, D. (1989). *A measure of uncertainty: The effects of the mass media* (Broadcasting Standards Council Research Monograph). London: John Libbey.

Cumberbatch, G., Lee, M., Hardy, G., & Jones, I. (1987). *The portrayal of violence on British television: A content analysis.* Birmingham, UK: Aston University, Applied Psychology Division.

Curr, W., Hallworth, H. J., & Wilkinson, A. M. (1962). How secondary modern school children spend their time. *Educational Review, 15,* 3–9.

Curr, W., Hallworth, H. J., & Wilkinson, A. M. (1964). Patterns of behavior in secondary modern school children. *Educational Review, 16,* 187–196.

Darcy, M. (1999). The discourse of "community" and the reinvention of social housing policy in Australia. *Urban Studies, 36,* 13–26.

Davies, N. (1997). *Dark heart: The shocking truth about Britain.* London: Chatto & Windus.

De Koning, T. L., Conradie, D. P., & Nel, E. M. (1990). The effect of different kinds of television programs on South African youth. *Advances in Learning and Behavioral Difficulties, 6,* 79–101.

Dorr, A., & Kunkel, D. (1990). Children and the media environment. *Communication Research, 17,* 5–25.

Downes, D., & Rock, P. (1995). *Understanding deviance: A guide to the sociology of crime and rule-breaking* (2nd ed., rev.). Oxford, England: Clarendon.

Edwards, A. (1990). *Fish and Fisheries of Saint Helena Island.* St. Helena, South Atlantic Ocean: Education Department.

Ekblad, S. (1990). The Children's Behavior Questionnaire for completion by parents and teachers in a Chinese sample. *Journal of Child Psychology and Psychiatry, 31,* 5, 775–791.

Elias, N. (1974). Foreword: Towards a theory of communities. In C. Bell & H. Newby (Eds.), *The sociology of community: A selection of readings.* London: Frank Cass.

Elliott, D., Wilson, W. J., Huizinga, D. R., Elliott, A., & Rankin, B. (1996). The effects of neighbourhood disadvantage on adolescent development. *Journal of Research in Crime and Delinquency, 33,* 389–426.

Emmett, I. (1982). Place, community and bilingualism in Blaenau Ffestiniog. In A. P. Cohen (Ed.), *Belonging: Identity and social organisation in British rural cultures.* Manchester, England: Manchester University Press.

Eron, L., Huesmann, R., Lefkowitz, M., & Walder, L. (1972). Does television violence cause aggression? *American Psychologist, 27,* 253–263.

Evans, D. (1994). *Schooling in the South Atlantic Islands, 1661–1992.* Oswestry, UK: Anthony Nelson.

Falk, J. (1998). The meaning of the web. *The Information Society, 14,* 285–293.

Farrell, M. P., & Barnes, G. M. (1993). Family systems and social support: A test of the effects of cohesion and adaptability on the functioning of parents and adolescents. *Journal of Marriage and the Family, 55,* 119–132.

Fenigstein, A. (1979). Does aggression cause a preference for viewing media violence? *Journal of Personality and Social Psychology, 37,* 2307–2317.

Foreign and Commonwealth Office (1999). *Partnership for Progress and Prosperity.* London: Her Majesty's Statistics Office.

Frankenberg, R. (1957). *Village on the border: A social study of religion, politics, and football in a North Wales community.* London: Cohen & West.

Freedman, J. L. (1984). Effect of television violence on aggressiveness. *Psychological Bulletin, 96,* 227–246.

Freedman, J. L. (1986). Television violence and aggression: A rejoinder. *Psychological Bulletin, 100,* 372–378.

French, J., & Pena, S. (1991). Children's hero play of the 20th century: Changes resulting from television's influences. *Child Study Journal, 21*, 79–94.

French, J., Pena, S., & Holmes, R. (1987). The superhero TV dilemma. *The Newsletter of Parenting, 10*, 8–9.

Friedrich-Cofer, L., & Huston, A. (1986). Television violence and aggression: The debate continues. *Psychological Bulletin, 100*, 364–371.

Furu, T. (1962). *Television and children's life: A before-after study*. Tokyo: Japan Broadcasting Corporation.

Gauntlett, D. (1995). *Moving experiences: Understanding television's influences and effects*. London: John Libbey.

Gauntlett, D. (1997). Why no clear answers on media effects? In T. Charlton & K. David (Eds.), *Elusive links: Television, video games and children's behaviour*. Tewkesbury, England: Park Published Papers.

Gerbner, G. (1972). Violence in television drama: Trends and symbolic functions. In G. A. Comstock & E. A. Rubinstein (Eds.), *Television and social behavior: Vol. 1. Media Content and Control* (pp. 28–187). Washington, DC: U.S. Government Printing Office.

Gerbner, G., & Gross, L. (1976). Living with television: The violence profile. *Journal of Communication, 26*, 173–199.

Gerbner, G., Gross, L., Jackson-Beeck, M., Jeffries-Fox, S., & Signorielli, N. (1978). Cultural indicators: Violence profile no. 9. *Journal of Communication, 28*, 176–207.

Gerbner, G., Gross, L., Morgan, M., & Signorielli, N. (1980). The "mainstreaming" of America: Violence profile no. 11. *Journal of Communication, 30*, 10–29.

Gerbner, G., Gross, L., Signorielli, N., Morgan, M., & Jackson-Beeck, M. (1979). The demonstration of power: Violence profile no. 10. *Journal of Communication, 29*, 177–196.

George, B. (1960). *Stories from the island of St. Helena*. St. Helena: Author.

George, B. (1994, September 16). Juvenile crime. *St. Helena News, 9*, 6–7.

George, B. (1998, September). *St. Helena—The Human Habitat: People, Places and Politics*. Paper presented at the meeting of the St. Helena Nature Conservation Group, St. Helena, South Atlantic Ocean.

Giddens, A. (1985). Time, space and regionalisation. In D. Gregory & J. Urry (Eds.), *Social relations and social structures*. London: Macmillan Education.

Gillett, A. N. (1979). *St. Helena: Schools for the future*. Unpublished manuscript.

Ginpil, S. (1976). Violent and dangerous acts on New Zealand television. *New Zealand Journal of Educational Studies, 10*, 152–157.

Goffman, E. (1969). *The presentation of self in everyday life*. London: Allen Lane.

Gordon, D. R., & Singer, P. D. (1977). Content analysis of the news media: Newspapers and television. *Report to the Royal Commission on violence in the communications industry: Vol. 3. Violence in television, films, and news* (pp. 482–494). Toronto, Canada: The Royal Commission.

Gosse, P. (1938). *St. Helena, 1502–1938*. London: Cassell.

Granzberg, G. (1982). Television as storyteller: The Algonkian Indians of central Canada. *Journal of Communication, 32*, 1, 43–52.

Granzberg, G. (1985). Television and self-concept formation in developing areas: The central Canadian Algonkian experience. *Journal of Cross-Cultural Psychology, 16*, 3, 313–328.

Granzberg, G., & Steinbring, J. (1980). *Television and the Canadian Indian* (Tech. Rep.). Winnipeg, Canada: University of Winnipeg, Department of Anthropology.

Green, J. P. (Ed.). (1982). *Social control: Views from the social sciences*. London: Sage.

Greenberg, B., & Gordon, T. F. (1972a). Children's perception of television violence: A replication. In G. A. Comstock, E. A. Rubinstein, & J. P. Murray (Eds.), *Television and social behavior: Vol. 5. Television's effects: Further explorations* (pp. 211–230). Rockville, MD: National Institute of Mental Health.

Greenberg, B., & Gordon, T. F. (1972b). Perceptions of violence in television programs: Critics and the public. In G. A. Comstock & E. A. Rubinstein (Eds.), *Television and social behavior: Vol. 1. Media content and control* (pp. 244–258). Rockville, MD: National Institute of Mental Health.

Greenberg, B., & Gordon, T. F. (1972c). Social class and racial differences in children's perceptions of television violence. In G. A. Comstock, E. A. Rubinstein, & J. P. Murray (Eds.), *Television and social behavior: Vol. 5. Television's effects: Further explorations* (pp. 185–210). Rockville, MD: National Institute of Mental Health.

Gunter, B. (1985). *Dimensions of television violence*. Aldershot, England: Gower.

Gunter, B. (1988). Editorial. *Current Psychology: Research & Reviews, 7*, 1, 3–9.

Gunter, B. (1998). *The effects of video games on children: The myth unmasked*. Sheffield, England: Sheffield Academic Press.

Gunter, B., & Charlton, T. (1999, May). *Violence on television in St. Helena: What has been happening?* Paper presented at the 49th Annual Conference of the International Communication Association, San Francisco, CA.

Gunter, B., Charlton, T., & Lovemore, T. (1998). Television on St. Helena: Does the output give cause for concern? *Medien Psychologie, 3*, 184–203.

Gunter, B., & Harrison, J. (1997a). Measuring television violence: Impressions given by different quantitative indicators. *Communications: European Journal of Communication Research, 21*, 385–406.

Gunter, B., & Harrison, J. (1997b). Violence in children's programs on British television. *Children and Society, 11*, 143–156.

Gunter, B., & Harrison, J. (1998). *Violence on television: A study of British programs*. London: Routledge.

Gunter, B., & McAleer, J. (1990). *Children and television: The one eyed monster?* London: Routledge.

Gunter, B., & McAleer, J. (1997). *Children and television* (2nd ed.). London: Routledge.

Gunter, B., & Stipp, H. (1992). Attitudes about sex and violence on television in the United States and Great Britain: A comparison of research findings. *Medien Psychologie, 44*, 267–286.

Gunter, B., & Ward, K. (1997). News reporting of television violence in the British press. *Medien Psychologie, 9*, 4, 253–270.

Gunter, B., & Wober, M. (1988). *Violence on television: What the viewers think*. London: John Libbey.

Haines, H. (1983). *Violence on television: A report on the Mental Health Foundation's media watch survey*. Auckland, New Zealand: Mental Health Foundation of New Zealand.

Hall, R. V. (1974). *The measurement of behavior*. Lawrence, KS: H & H Enterprises.

Halloran, J. D., & Croll, P. (1972). Television programs in Great Britain: Content and control. In G. A. Comstock & E. A. Rubinstein (Eds.), *Television and social behavior: Vol. 1. Media content and control* (pp. 415–492). Washington, DC: U.S. Government Printing Office.

Hannan, A. (1999, May). *A clash of cultures? How the children of St. Helena responded to the introduction of television with specific reference to diary-based surveys of leisure activities*. Paper presented at the 49th Annual Conference of the International Communication Association, San Francisco, CA.

Hannan, A., & Charlton, T. (1999). Leisure activities of middle school pupils of St. Helena before and after the introduction of television. *Research Papers in Education, 14*, 257–274.

Harper, S. (1989). The British rural community: An overview of perspectives. *Journal of Rural Studies, 5*, 161–184.

Hawtin, M., Hughes, G., & Percey-Smith, J. (1994). *Community profiling: Auditing social needs*. Buckingham, England: Open University Press.

Himmelweit, H. T., Oppenheim, A. N., & Vince, P. (1958). *Television and the child: An empirical study of the effect of television on the young*. London: Oxford University Press.

Hirschi, T. (1969). *Causes of delinquency*. Berkeley, CA: University of California Press.

Hodge, R., & Tripp, D. (1985). *Children and television*. Cambridge, England: Polity Press.

Hodge, R., & Tripp, D. (1986). *Children and television: A semiotic approach*. Cambridge, England: Polity Press.

Hoffner, C. (1996, June). *News media framing of the television violence issue*. Paper presented at the Duke University Conference on Media Violence and Public Policy in the Media, Durham, NC.

Holsti, O. R. (1969). *Content analysis for the social sciences and humanities*. Reading, MA: Addison-Wesley.

Holy, L., & Stuchlik, M. (1983). *Actions, norms and representations: Foundations of anthropological inquiry*. Cambridge, England: Cambridge University Press.

Huesmann, L. R. (1986). Psychological processes promoting the relation between exposure to media violence and aggressive behaviour by the viewer. *Journal of Social Issues, 42*, 125–140.

Huesmann, L. R., & Eron, L. D. (Eds.). (1986). *Television and the aggressive child: A cross-national comparison*. Hillsdale, NJ: Lawrence Erlbaum Associates.

Institute for Cooperative Relations [ISCORE] (1999). *The ICS research program*. <http://www.fss.uu.nl/soc/iscore>, accessed 9.3.1999.

Iwao, S., de Sola Pool, I., & Hagiwara, S. (1981). Japanese and US media: Some cross-cultural insights into TV violence. *Journal of Communication, 31*, 2, 28–36.

Jacob, S., Bourke, L., & Luloff, A. E. (1997). Rural community stress, distress, and well-being in Pennsylvania. *Journal of Rural Studies, 13*, 275–288.

Jennings, C. M., & Gillis-Olion, M. (1980). *The impact of television cartoons on child behaviour*. Paper presented at the Annual Meeting of the National Association for the Education of Young Children, Atlanta, GA. (ERIC Document Reproduction Service No. 194-184)

Jenson, E., & Graham, E. (1995). Addressing media violence: An overview. In C. Wekessar (Ed.), *Violence in the media*. San Diego, CA: Greenhaven Press.

Jo, E., & Berkowitz, L. (1994). A priming effect analysis of media influences: An update. In J. Bryant & D. Zillman (Eds.), *Media effects: Advances in theory and research*. Hillsdale, NJ: Lawrence Erlbaum Associates.

Jones, G. (1999). "The same people in the same places?": Socio-spatial identities and migration in youth. *Sociology, 33*, 1–22.

Jones, K., Charlton, T., & Wilkins, J. (1995). Classroom behaviours which first and middle school teachers in St. Helena find troublesome. *Educational Studies, 21*, 139–153.

Joy, L. A., Kimball, M. M., & Zabrack, M. M. (1986). Television and children's aggressive behavior. In T. M. Williams (Ed.), *The impact of television: A natural experiment in three settings*. New York: Academic Press.

Kastrup, M. (1976). Psychic disorders among pre-school children in a geographically delimited area of Aarhus County, Denmark. *Acta Psychiatrica Scandinavica, 54*, 29–42.

Kiesler, C. A., & Kiesler, S. B. (1969). *Conformity*. London: Addison-Wesley.

Kostelnik, M. J., Whiren, A. P., & Stein, L. C. (1986). Living with he-man—managing superhero fantasy play. *Young Children, 41*, 4, 3–9.

Krcmar, M., & Greene, K. (1999). Predicting exposure to and uses of television violence. *Journal of Communication, 49*, 3, 24–45.

Krippendorf, K. (1980). *Content analysis: An introduction to its methodology*. Newbury Park, CA: Sage.

Kunkel, D., Wilson, B. J., Linz, D., Potter, J., Donnerstein, E., Smith, S. L., Blumenthal, E., & Gray, T. (1996, June). *Content analysis of entertainment television: Implications for public policy*. Paper presented at the Duke University Conference on Media Violence and Public Policy in the Media, Durham, NC.

Lambert, R. (1968). *The hothouse society: an exploration of boarding-school life through the boys' and girls' own writings*. London: Weidenfeld and Nicolson.

Langellier, K., & Peterson, E. E. (1993). Family storytelling as a strategy of social control. In D. K. Murphy (Ed.), *Narrative and social control: Critical perspectives*. London: Sage.

Laub, J. H., & Lauritsen, J. L. (1998). The interdependence of school violence with neighborhood and family conditions. In D. S. Elliott, B. Hamburg, & K. R. Williams (Eds.), *Violence in American schools: A new perspective*. Cambridge, England: Cambridge University Press.

Liebert, R. M., & Baron, R. A. (1973). Some immediate effects of televised violence on children's behaviour. *Developmental Psychology, 6*, 469–475.

Lin, C. A. (1992). The functions of the VCR in the home leisure environment. *Journal of Broadcasting and Electronic Media, 36*, 345–351.

Linden, R., Currie, R. F., & Driedger, L. (1985). Interpersonal ties and alcohol use among Mennonites. *Canadian Review of Sociology and Anthropology, 22*, 559–573.

Linton, J. M., & Jowitt, G. S. A content analysis of feature films. *Report to the Royal Commission on violence in the communications industry: Vol. 3. Violence in television, films, and news* (pp. 574–580). Toronto, Canada: The Royal Commission.

Little, J., & Austin, P. (1996). Women and the rural idyll. *Journal of Rural Studies, 12*, 101–111.

Luloff, A. E., & Swanson, L. E. (1990). *American rural communities*. Boulder, CO: Westview Press.

Lynn, R., Hampson, S., & Agahi, E. (1989). Television violence and aggression: A genotype-environment, correlation and interaction theory. *Social Behavior and Personality, 17*, 2, 143–164.

Maccoby, E. E., & Wilson, W. C. (1957). Identification and observational learning from films. *Journal of Abnormal and Social Psychology, 55*, 76–87.

Matsuura, M., Okubo, Y., Kojima, T., Takahashi, R., Wang, Y. F., Shen, Y. C., & Lee, C. K. (1993). A cross-national prevalence study of children with emotional and behavioral problems—A WHO collaborative study in the Western Pacific Region. *Journal of Child Psychology and Psychiatry, 34*, 3, 307–315.

McCann, T. E., & Sheehan, P. W. (1985). Violence content in Australian television. *Australian Psychologist, 20*, 33–42.

McFarlane, G. (1977). Gossip and social relations in a Northern Irish community. In M. Stuchlik (Ed.), *Goals and behaviour*. Belfast, UK: The Queens University of Belfast.

McFarlane, G. (1986). "It's not as simple as that": The expression of the Catholic and Protestant boundary in Northern Irish rural communities. In A. P. Cohen (Ed.), *Symbolising boundaries: Identity and diversity in British cultures*. Manchester, England: Manchester University Press.

McGhee, R., Silva, P. A., & Williams, S. (1984). Behavior problems in a population of seven-year-old children: Prevalence, stability and types of disorder—A research report. *Journal of Child Psychology and Psychiatry, 25*, 251–259.

McGilvery, L. (1991, September). *Living happily with television*. Paper presented at the Early Childhood Conference, Dunedin, New Zealand.

McGuire, J., & Richman, N. (1986). Screening for the behaviour problems in nurseries: The reliability and validity of the pre-school behaviour checklist. *Journal of Child Psychology and Psychiatry, 27*, 7–32.

McGuire, J., & Richman, N. (1988). *The Pre-School Behaviour Checklist*. Windsor, UK: NFER-Nelson.

McIntyre, J. J., Teevan, Jr., J. J., & Hartnagel, T. (1972). Television violence and deviant behavior. In G. A. Comstock & E. A. Rubinstein (Eds.), *Television and social behavior: Vol. 3. Television and adolescent aggressiveness* (pp. 173–238). Washington, DC: U.S. Government Printing Office.

Menon, V. (1993). Violence on television: Asian data for an Asian standard. *Intermedia, 40,* 515–543.

Merrett, F., & Wheldall, K. (1986). Observations of pupils and teachers in classrooms (OPTIC): A behavioural observation schedule for use in schools. *Educational Psychology, 6,* 57–70.

Milavsky, J. R., Kessler, R., Stipp, H., & Rubens, W. S. (1982). *Television and aggression: A panel study.* New York: Academic Press.

Milgram, S., & Shotland, R. L. (1973). *Television and antisocial behavior: Field experiments.* New York: Academic Press.

Mischel, W. (1970). Sex differences. In P. Mussen (Ed.), *Carmichael's manual of child psychology: Vol. 2.* New York: Wiley.

Morley, D. (1992). *Television, audiences and cultural studies.* London: Routledge.

Murray, J. P., & Kippax, S. (1978). Children's social behavior in three towns with differing television experience. *Journal of Communication, 30,* 19–29.

Mussen, P. H., Conger, J. J., & Kagan, J. (1969). *Child development and personality.* New York: Harper & Row.

Mutz, D. C., Roberts, D. F., & Van Vuuren, D. P. (1993). Reconsidering the displacement hypothesis: Television's influence on children's time use. *Communication Research, 20,* 51–75.

National Commission on the Causes and Prevention of Violence. (1969). *To Establish Justice, To Insure Domestic Tranquility: Final Report of the National Commission on the Causes and Prevention of Violence.* Washington, DC: U.S. Government Printing Office.

Newell, D., & Shaw, I. (1972). *Violence on television: Program content and viewer perception.* London: British Broadcasting Corporation.

Office for National Statistics. (1998). Lifestyles. *Social Trends, 28,* 216.

O'Neal, E. C., & Taylor, S. L. (1989). Status of the provoker, opportunity to retaliate, and interest in video violence. *Aggressive Behavior, 15,* 171–180.

Overseas Development Administration. (1993). *Report on sustainable environment and development strategy and action plan for St. Helena* (Vols. 1–2). London: Author.

Panting, C., Coles, D., & Abrahams, M. (1999, May). *Broadcast television viewing habits and social behavior of 7- to 8-year-olds on St. Helena.* Paper presented at the annual conference of the International Communication Association, San Francisco, CA.

Patterson, G. (1980). Children who steal. In T. Hersch & M. Gottfredson (Eds.), *Understanding crime: Current theory and research* (pp. 73–90). Beverley Hills, CA: Sage.

Pearl, D., Bouthilet, L., & Lazar, J. (Eds.). (1982). *Television and behaviour: Ten years of scientific progress and implications for the eighties: Vol. 2.* Washington, DC: U.S. Government Printing Office. Rockville, MD: National Institute of Mental Health.

Pepler, D. J., & Craig, W. M. (1995). A peek behind the fence: Naturalistic observations of aggressive children with remote audiovisual recording. *Development Psychology, 31,* 548–553.

Phillips, S. K. (1986). Natives and incomers: The symbolism of belonging in Muker Parish, North Yorkshire. In A. P. Cohen (Ed.), *Symbolising boundaries: Identity and diversity in British cultures.* Manchester, England: Manchester University Press.

Potter, J., Linz, D., Wilson, B. J., Kunkel, D., Donnerstein, E., Smith, S. L., Blumenthal, E., & Gray, T. (1996, June). *Content analysis of entertainment television: New methodological developments.* Paper presented at the Duke University Conference on Media Violence and Public Policy in the Media, Durham, NC.

Potts, R., Huston, A. C., & Wright, J. C. (1986). The effects of television form and violent content on boys' attention and social behavior. *Journal of Experimental Child Psychology, 41,* 1–17.

Raub, W., & Weesie, J. (1990). Reputation and efficiency in social interactions: An example of network effects. *American Journal of Sociology, 96,* 626–654.

Reeves, B. (1979). Children's understanding of television people. In E. Wartella (Ed.), *Children communicating: Media and development of thought, speech, understanding.* Beverly Hills, CA: Sage.

Reiss, A. J. (1995). Community influences on adolescent behaviour. In M. Rutter (Ed.), *Psychosocial disturbances in young people: Challenges for prevention* (pp. 305–332). New York: Cambridge University Press.

Remmers, H. H. (1954). *Four years of New York television: 1951–1954.* Urbana, IL: National Association of Educational Broadcasters.

Richman, N., & Graham, P. J. (1971). A behavioral screening questionnaire for use with 3-year-old children. Preliminary findings. *Journal of Child Psychology and Psychiatry, 12,* 5–33.

Roberts, D. F., Henriksen, L., Voelker, D. H., & Van Vuuren, D. P. (1993). Television and schooling: Displacement and distraction hypotheses. *Australian Journal of Education, 37,* 198–211.

Rotter, J. B. (1966). Generalized expectancies for internal versus external control of reinforcement. *Psychological Monographs, 80* (Whole No. 609).

Royle, S. A. (1982). Attitudes and aspirations on St. Helena in the face of continued economic dependency. *The Geographical Journal, 158,* 31–39.

Rubin, A. M., & Bantz, C. R. (1987). Utility of videocassette recorders. *American Behavioral Scientist, 30,* 471–485.

Rutter, M. (1967). A children's behaviour questionnaire for completion by teachers: Preliminary findings. *Journal of Child Psychology and Psychiatry, 8,* 1–11.

Rutter, M., Cox, A., Tupling, C., Berger, M., & Yule, W. (1975). Attainment and adjustment in two geographical areas. *British Journal of Psychiatry, 126,* 493–509.

Rutter, M., Tizard, J., & Whitmore, K. (1970). *Education, health and behaviour.* London: Longman.

St. Helena Development and Economic Planning Department. (1998). *St. Helena Census.* St. Helena: Author.

St. Helena Diocesan Magazine (1925), 24.

St. Helena Diocesan Magazine (1927). XXVIII, 312, November, 61.

St. Helena Diocesan Magazine (1929). XXVIII, 330, May, 61.

St. Helena Government (1980). St. Helena development survey—economic considerations, adapted from papers submitted by J. A. Barnett.

St. Helena Government (1997a). *The St. Helena Police Force public survey.*

St. Helena Government (1997b). *Police statistics: Period 1st January–31st December 1997.*

St. Helena News (1995). *Launch of the St. Helena Television Service, 9, 37.*

Saunders, P., Newby, H., Bell, C., & Rose, D. (1978). Rural community and rural community power. In H. Newby (Ed.), *International perspectives on rural sociology.* Chichester, England: Wiley.

Sawin, D. (1990). Aggressive behaviour among children in small playground settings with violent television. *Advances in Learning and Behavior Disabilities, 6,* 157–177.

Schramm, W., Lyle, J., & Parker, E. (1961). *Television and the lives of our children.* Stanford, CA: Stanford University Press.

Schulenburg, A. H. (1998). St. Helena: British local history in the context of empire. *The Local Historian, 28,* 108–122.

Schulenburg, A. H. (in press). "Island of the blessed": Eden, Arcadia and the picturesque in the textualizing of St. Helena. *Journal of Historical Geography.*

Schulenburg, A. H. (n.d.). 'Friendly', 'caring', 'nosey': Secondary school pupils' perceptions of 'People on St. Helena' (unpublished manuscript).

Scott, J. (1990). *Domination and the arts of resistance: Hidden transcripts.* London: Yale University Press.

Scott, S. (1996). Measuring oppositional and aggressive behaviour. *Child Psychology and Psychiatry, Review, 1,* 104–109.

Shinar, D., Parnes, P., & Caspi, D. (1972). Structure and content of television broadcasting in Israel. In G. A. Comstock & E. A. Rubinstein (Eds.), *Television and behavior: Ten years of scientific progress and implications for the eighties* (Vol. 1). Washington, DC: U.S. Government Printing Office.

Signorielli, N., Gross, L., & Morgan, M. (1982). Violence in television programs: Ten years later. In D. Pearl, L. Bouthilet, & J. Lazar (Eds.), *Television and behavior: Ten years of scientific progress and implications for the eighties* (Vol. 2) (pp. 158–173). Washington, DC: U.S. Government Printing Office.

Singer, D. G., & Singer, J. L. (1980). Television viewing and aggressive behavior in preschool children: A field study. *Annals of the New York Academy of Science, 347,* 289–303.

Singer, D. G., Singer, J. L., & Zuckerman, D. M. (1990). *Use TV to your child's advantage.* Reston, VA: Acropolis Books.

Singer, J. L., & Singer, D. G. (1986a). Family experiences and television viewing as predictors of children's imagination, restlessness, and aggression. *Journal of Social Issues, 42,* 107–124.

Singer, J. L., & Singer, D. G. (1986b). Television viewing and family communication style as predictors of children's emotional behavior. *Journal of Children in Contemporary Society, 17,* 75–91.

Sluckin, A. (1981). *Growing up in the playground.* London: Routledge.

Smith, S. L., Nathanson, A. I., & Wilson, B. (1999, May). *Violence in prime time: An analysis of the amount and context of aggression.* Paper presented at the Annual Conference of the International Communications Association, San Francisco, CA.

Smythe, D. W. (1954). *Three years of New York television: 1951–1953.* Urbana, IL: National Association of Educational Broadcasters.

Sprafkin, J., Gadow, K. D., & Abelman, R. (1992). *Television and the exceptional child: A forgotten audience.* Hillsdale, NJ: Lawrence Erlbaum Associates.

Stewart, D. E. (1983). *The television family: A content analysis of the portrayal of family life in prime time television.* Melbourne, Australia: Institute of Family Studies.

Stipp, H., & Milavsky, J. R. (1988). US television programming's effects on aggressive behaviour of children and adolescents. *Current Psychology: Research & Reviews, 7,* 76–92.

Sun, L. (1989). *Limits of selective viewing: An analysis of "diversity" in dramatized programming.* Unpublished master's thesis, University of Pennsylvania, Philadelphia.

Surgeon General's Scientific Advisory Committee on Television and Social Behavior. (1972). *Television and growing up: The impact of televised violence. Report to the surgeon general, United States public health service.* Washington, DC: U.S. Government Printing Office.

Tittle, C. R. (1980). *Sanctions and social deviance: The question of deterrence.* New York: Praeger.

Tomlinson, J. (1997). Cultural globalization and cultural imperialism. In A. Mohammadi (Ed.), *International communication and globalization: A critical introduction.* London: Sage.

Truglio, R. T., Murphy, K. C., Oppenheimer, S., Huston, A. C., & Wright, J. C. (1996). Predictors of children's entertainment television viewing: Why are they tuning in? *Journal of Applied Developmental Psychology, 17,* 475–493.

United Nations Development Programme. (1999). *Human development report for St. Helena.* New York: Author.

Valentine, G. (1997). A safe place to grow up?: Parenting, perceptions of children's safety and the rural idyll. *Journal of Rural Studies, 13,* 137–148.

Van der Vort, T. H. A. *Television violence: A child's eye view.* Amsterdam: Elsevier.

Van Evra, J. (1998). *Television and child development.* Hillsdale, NJ: Lawrence Erlbaum Associates.

von Feilitzen, C. (1999). Children's amount of TV viewing. In C. Feilitzen & U. Carlsson (Eds.), *Children and media: Image, education, participation*. Goteborg, Sweden: UNESCO.

Ward, F. (1959). Letter. *The St. Helena "Wirebird", 2*, 430.

Watson, C., Bassett, G., Lambourne, R., & Shuker, R. (1991). *Television violence: An analysis of the portrayal of "violent acts" on the three New Zealand broadcast television channels during the week of 11th–17th February 1991*. Palmerston North, New Zealand: Massey University, Educational Research and Development Centre.

Wiegman, O., Kuttschreuter, M., & Baarda, B. (1992). A longitudinal study of the effects of television viewing on aggressive and pro-social behaviours. *British Journal of Social Psychology, 31*, 147–164.

Wigglesworth, A., & Vulliamy, E. (1997, April 15). Far-flung British island revolts. *The Guardian*, 1–2.

Williams, T. M. (Ed.). (1986). *The impact of television: A natural experiment in three communities*. New York: Academic Press.

Williams, T. M., & Boyes, M. C. (1986). Television-viewing patterns and use of other media. In T. M. Williams (Ed.), *The impact of television: A natural experiment in three communities* (pp. 215–263). New York: Academic Press.

Williams, T. M., & Handford, A. G. (1986). Television and other leisure activities. In T. M. Williams (Ed.), *The impact of television: A natural experiment in three communities* (pp. 143–213). London: Academic Press.

Williams, T. M., Zabrack, M. L., & Joy, L. A. (1982). The portrayal of aggression on North American television. *Journal of Applied Social Psychology, 12*, 5, 360–380.

Wilson, B. J., Donnerstein, E., Linz, D., Kunkel, D., Potter, J., Smith, S. L., Blumenthal, E., & Gray, T. (1996, June). *Content analysis of entertainment television: The importance of context*. Paper presented at the Duke University Conference on Media Violence and Public Policy in the Media, Durham, NC.

Wilson, B. J., Kunkel, D., Linz, D., Potter, J., Donnerstein, E., Smith, S. L., Blumenthal, E., Berry, M., & Federman, J. (1998). The nature and context of violence on American television. In U. Carlsson & C. von Feilitzen (Eds.), *Children and media violence*. Goteborg, Sweden: UNESCO.

Wilson, B. J., Smith, S., Linz, D., Potter, J., Donnerstein, E., Kunkel, D., Blumenthal, E., & Gray, T. (1996, June). *Content analysis of entertainment television: The 1994–1995 results*. Paper presented at the Duke University Conference on Media Violence and Public Policy in the Media, Durham, NC.

Wood, W., Wong, F., & Chachere, J. (1991). Effects of media violence on viewers' aggression in unconstrained social interaction. *Psychological Bulletin, 109*, 371–383.

Wright, S. (1992). Image and analysis: new directions in community studies. In B. Short (Ed.), *The English rural community: Image and analysis*. Cambridge, England: Cambridge University Press.

Young, E. D. K. (1986). Where the daffodils blow: Elements of communal imagery in a northern suburb. In A. P. Cohen (Ed.), *Symbolising boundaries: Identity and diversity in British cultures*. Manchester, England: Manchester University Press.

Zekeri, A. A., Wilkinson, K. P., & Humphrey, C. R. (1994). Past activeness, solidarity, and local development efforts. *Rural Sociology, 59*, 216–235.

Zillmann, D. (1991). Television viewing and physiological arousal. In J. Bryant & D. Zillmann (Eds.), *Responding to the screen: Reception and reaction processes* (pp. 103–133). Hillsdale, NJ: Lawrence Erlbaum Associates.

Zillmann, D., & Bryant, J. (1991). Responding to comedy: The sense and nonsense in humor. In J. Bryant & D. Zillmann (Eds.), *Responding to the screen: Reception and reaction processes* (pp. 261–279). Hillsdale, NJ: Lawrence Erlbaum Associates.

Author Index

Subject Index